WOMEN'S LIFEWORLDS

Women's Lifeworlds aims to make explicit the diversity and complexity of women's perceptions and reactions to their own lifeworlds in their own words. It highlights women's personal perceptions of the basic dimensions of their own lives, their sources of strength and the things that bring meaning to their lives.

Fifteen women of various age groups and from different cultural, religious, social and geographical backgrounds – from Mexican politician, Muslim psychiatrist, Finnish housewife to Indian guru and African rural woman - write about themselves, the lives of their grandmothers, their mothers and their daughters or other women of those generations close to the family. These life narratives serve to prepare an encounter at which the authors come together to share their perceptions of the currents and undercurrents of the narratives. They tell the stories of fifty-four women in Asia, Africa, Latin America and Europe, covering a period of more than a hundred years.

These engaging narratives illustrate the changing meaning of 'place' in women's lives over time and across space. They move beyond traditional feminist concerns and North–South divides. They highlight women's own implicit ways in day-to-day dealings with their lives of disregarding the dominant tendency to define realities in terms of dichotomies. These ways intertwine rather than separate seemingly opposite elements, and the analysis creates an awareness of misleading similarities created by the use of dominant languages in intercultural relations.

Edith Sizoo is International Coordinator for South North Network Cultures and Development in Brussels.

INTERNATIONAL STUDIES OF WOMEN AND PLACE

Edited by Janet Momsen, *University of California at Davis*
and Janice Monk, *University of Arizona*

The Routledge series of *International Studies of Women and Place* describes the diversity and complexity of women's experience around the world, working across different geographies to explore the processes which underlie the construction of gender and the lifeworlds of women.

WOMEN'S LIFEWORLDS

Women's narratives on shaping their realities

Edited by Edith Sizoo

London and New York

First published 1997
by Routledge
11 New Fetter Lane, London EC4P 4EE

Simultaneously published in the USA and Canada
by Routledge
29 West 35th Street, New York, NY 10001

© 1997 Edith Sizoo

Typeset in Baskerville by Routledge
Printed and bound in Great Britain by TJ International Ltd,
Padstow, Cornwall

British Library Cataloguing in Publication Data
A catalogue record for this book is available from the British Library

Library of Congress Cataloguing in Publication Data
A catalogue record has been requested

ISBN 0–415–17176–8 (hbk)
ISBN 0–415–17177–6 (pbk)

CONTENTS

Part III The findings

CONTRIBUTORS

Esperanza Abellana, born in 1947 in the Philippines, has played an active role in organizing women in rural and urban poor communities. She served the Asian Social Institute in Manila as its Academic Coordinator, the Center for Asian Dialogue and a Center for Spirituality before switching to the Philippine Office of the Save the Children Fund.

Durre Sameen Ahmed, born in 1949 in Pakistan, is Professor of Psychology and Communication at the National College of Arts, Lahore; she is also a practising psychologist. Her interests include issues pertaining to the psychology of women, development, religion and the media.

Jaana Airaksinen, born in 1961 in Finland, worked for the United Nations Development Programme in Papua New Guinea and Malawi as an economist. She is involved in the Third World solidarity movement and is an active feminist.

Ethel Crowley, born in 1968 in Ireland, teaches at the Sociology Department of University College Cork. She has carried out research in the area of feminism, with regard to Ireland as well as the Middle East and Latin America. She is currently conducting research on the impact of EU agricultural policy on rural Ireland.

Yvonne Deutsch, born in Timisoara, Romania, came to Israel with her mother when she was 8 years old. She studied Hebrew Literature, African History and Social Work. As a peace activist she was one of the initiators of the Women in Black movement opposing the Israeli occupation of Palestinian territories. She currently runs a women's centre in Tel Aviv.

Kamala Ganesh, born in 1952 in South India, is Reader in Cultural Anthropology at the University of Bombay. From 1988 to 1993 she was the Secretary of the Commission on Women of the International Union of Anthropological and Ethnological Sciences. Her interests have been in the area of the interface of gender with caste and kinship in South Asia.

Shanti George, born in London in 1954, grew up in various parts of India, where she studied Social Anthropology. She has been Reader at the University of

Delhi, Visiting Senior Lecturer at the University of Zimbabwe, and is now based in The Netherlands. Her research includes dairy policies and practice in India and Zimbabwe.

Christina Gualinga was born in 1962 in a small village in the Equatorial Amazon (Ecuador). After secondary school, she became involved in the Indigenous movement. Married to a Belgian development worker, she is now active in Belgium in solidarity work for Indigenous peoples, linking up European solidarity groups and Indigenous movements in Latin America.

Amal Krieshe, born in 1957 on the West Bank in Palestine, studied Psychology in Jordan. She was member of the central committee of the former Communist Party and established the Palestinian Working Women's Society. She is also one of the founders of a Palestinian–Israeli women's network which strives to improve relations between Palestinian and Israeli women.

Nicole Note, born in Flanders in 1959, studied Indian American Cultures and Languages at the University of Leiden (The Netherlands), and served as a research assistant at the South North Network Cultures and Development. She is currently working as an educational staff member at the University Centre for Development Cooperation (UCOS) in Brussels.

Eliane Pontiguara, born in Brazil, was named one of the top ten women of 1988 by the National Council of Women for Brazil, for her contribution to 'the integration of women in the process of social, political and economic development in the country'. She is regional coordinator of the Indigenous Nations' Union and co-founder of the Women's Group for Indigenous Education.

Dolores Rojas Rubio, born in Mexico City in 1963, studied Electronic Engineering, Theatre and Political Analysis. She worked with NGOs involved in refugee aid and as Media Coordinator in the electoral campaign of the Democratic Revolution Party. She is currently Technical Secretary in the Legal Issues and State Reform Secretariat of this party.

Safia Mohamed Safwat, born in the Sudan of an Egyptian mother and a Sudanese father, became the second woman to join the chambers of the Attorney General of Sudan. Political restrictions forced her to settle in London. She has been actively involved in human rights and runs her own law practice. She also teaches Islamic Law and Human Rights at the University of London.

Safiatu Kassim Singhateh, born in The Gambia, is an international consultant in women and development. She coordinated the African NGOs' preparatory processes for the World Conference on Women in Beijing, 1995. She is currently working with the African Women's Development and Communication Network, FEMNET, as the Acting Executive Director.

Edith Sizoo, born in 1939 in The Netherlands, worked for seven years in Hong Kong and India. She directed a Netherlands umbrella organization of development NGOs and is now international coordinator at the South North Network Cultures and Development. Her research is focusing on a sociolinguistic approach to intercultural relations.

PREFACE
Women's ways of shaping their realities

This book aims to make explicit the diversity and complexity of women's perceptions and reactions to their own lifeworlds in their own words.

Fifteen women of various age groups, from different cultural, religious, social and geographical backgrounds, were asked to write about the lives of their grandmother, their mother, themselves and their daughters (or, if this was impossible, other women of those generations close to the family). These life narratives served to prepare an encounter at which the authors came together to share their perceptions of the currents and undercurrents of the narratives. They tell the stories of fifty-four women in Asia, Africa, Latin America and Europe, covering a period of more than a hundred years.

The book consists of three main Parts: an introduction on the theme and the process which led to the book; the life narratives themselves; and the findings of the analysis of the stories.

The insights gained from the analysis strengthen the motivation for the theme, in the sense that they confirm the need for a nuanced and specific approach to women's perceptions and ways of shaping their realities. The narratives illustrate the changing meaning of 'place' in women's lives over time and across space. They have a strong de-stereotyping effect. They move beyond traditional feminist concerns and North–South divides. They highlight women's own implicit ways, in day-to-day dealings with their lives, of disregarding the dominant tendency to define realities in terms of dichotomies. These ways intertwine rather than separate seemingly opposite elements. And – last but not least – the analysis makes one aware of misleading similarities created by the use of dominant languages in intercultural relations.

Hopefully the reader will discover for herself/himself other unveiled aspects.

For stories tell different things to different people . . .

ACKNOWLEDGEMENTS

The initiative to start a process of cross-cultural reflection on women's ways of shaping their realities was taken by the Regional Base in Europe of the South North Network Cultures and Development (hereafter 'Network Cultures'). This worldwide network consists of people involved in development thinking and practice, working in non-governmental organizations, people's movements and universities in all continents. Network Cultures tries to stimulate people at local levels to take stock of their own experiences, to share and reflect on them together with those who are in the midst of comparable experiences, and to disseminate the results of these analyses through publications and teaching. Network Cultures is conducting various thematic programmes, including Cultures and the Economy, Cultures and Conflict, Cultural Dynamics in Social Transformation, Arts and Culture in Development Education, Cultures and Women, and Methodology for Culture-Sensitive Approaches to Development.

The responsibility for the thematic programme Cultures and Women was entrusted to Edith Sizoo, who initiated the process which led to the production of this book. Within this initiative, Nicole Note provided invaluable assistance with the preparation and organization of the Encounter of the fifteen authors and the analysis of the findings. After this event she produced a record of the discussions. This exercise highlighted to what extent the Encounter proceedings made for additional inputs and contributed to a better understanding of the central questions raised by the theme. Faithful and competent secretarial support was given by Nicole Fraeys.

An intercultural Core Group helped to monitor the process. This group consisted of Shanti George (India/The Netherlands), Safia Safwat (Sudan), Ethel Crowley (Ireland), Marie-Claire Foblets (Belgium), Farida Sheriff (Zanzibar), Nicole Note (Belgium) and Edith Sizoo (The Netherlands).

Most relevant comments on the first draft of Parts I and III of this book were received from the workshop participants – in particular Jaana Airaksinen and Shanti George, and from one outsider, Sipko de Boer.

Thanks to the challenging suggestions from the two series editors, Janice Monk and Janet Momsen, the initial analysis was enriched with reflections on the role of 'place' in women's lives. Their insightful comments and Routledge editor Sarah Lloyd's encouraging guidance have made the publication of this book possible.

English-language editing was provided by Paula Bownas, with a view to improving the readability of some of the texts.

Financially the process was made possible by the Commission of the European Communities and the NGO donor agencies supporting Network Cultures Europe, while special grants came from Cebemo (The Netherlands), Fastenopfer (Switzerland) and the Finnish Ministry of Foreign Affairs.

To each and all of these actors we extend our warmest thanks.

Edith Sizoo
Editor
Brussels, December 1996

ILLUSTRATIONS

Cover photo and photos for Chapters 2, 5 and 7–16: courtesy Jan Stegeman

Photo for Chapter 3: courtesy J. Danois/NCOS Archives

Photo for Chapter 4: courtesy P. Van Wouwe/NCOS Archives

Photo for Chapter 6: courtesy K. Maes/NCOS Archives

Part I

INTRODUCTION

1

A POLYLOGUE

Edith Sizoo

The Encounter turned into a real-life meeting. Having come with pen and paper, I found myself unable to write down the very secret of reciprocity that in my view characterized our exchanges, the secret that true dialogue between people and cultures depends upon.

(a participant)

EXPLORING SIMILARITIES AND DIFFERENCES

Adding stones to a historical path

Over the last few decades tremendous efforts have been made to reach a better understanding of the causes of women's problems in society and to set out strategies to solve them. For many women the rising awareness of worldwide female solidarity was a vital spark in their personal life, igniting consciousness of their own specific reality.

Slowly but surely women started to speak out, publicly and privately, wanting to be heard openly and to be listened to on their own – possibly differing – terms. Although women's feelings about being a woman within their own societies were perhaps voiced most eloquently through feminist movements, less formally organized and less articulate women also had their own perceptions of themselves and their societies.

Feminist thinking and acting, outspoken as it was, may have created the impression of dealing with one universal problem of submission to patriarchal models of society, and of working on global, uniform strategies to empower women. However, a closer look at the emphases placed on the various aspects of the issue shows not only different waves in feminist thinking but also its inherent heterogeneity. We cannot speak of one feminism. There exists a variety of feminisms, each with its own angles and foci.[1]

At different times and at different places, one finds varying degrees of emphasis on such issues as women's rights in society[2], emancipatory politics and life politics.[3] Many feminists focus particularly on gender equality and equal access to decision-making,[4] or on the 'relations of ruling'.[5] Some, especially those from the southern hemisphere, feel that feminists should not be too preoccupied with sexuality. They challenge the hierarchies within the global feminist sphere and denounce Western feminist writing as 'colonial discourse' on '*the* Third World woman', thus ignoring differences.[6] At the same time, there are many Northern feminisms which acknowledge or advocate plurality.[7] Sometimes they go even further by challenging the whole category of 'woman' as too fixed and too dichotomized, while life in fact is much more fluid.[8] For some within this heterogeneous feminist movement, explicit political action through a well-organized movement is the main focus.[9] For others, it is not even on the agenda.

Solidarity and differences

After the recent decades of worldwide discussion on 'women's issues', we now seem to have entered a period of exploring the ways in which women respond to their immediate, and often very different, environments. The focus of attention is moving towards discovering the nature and quality of differences and the characteristics of their meaning.

Similarities and differences are explored within the categories of woman[10] as well as the social construction of woman/man and the consequences of this for the dealings that each have with their environment.[11,12] In this sense there is a shift towards defining and explaining differences in order to broaden our understanding of the contribution they can make. We could even speak of a move away from a focus on womanhood, towards an understanding of differences in general.

This tendency may bring feminist thinking closer to those – both women and men – who have been distanced in the past by a lack of subtlety and nuance. It may even bring it closer to feminists themselves, for even women who have been active in women's movements recognize that in their own lives there are obvious differences between the public stands they take on feminism or gender issues, on the one hand, and shifts or compromises which they make in their own lives as women, friends, lovers and mothers on the other. As so often happens, in many realms of life, we may hesitate to reveal the discrepancies between what we say, do and write, and what we live – but they do exist.

The NGO International Forum on Women, in conjunction with the United Nations Conference on Women and Development in Beijing, China (1995), has clearly shown that feminist thinking has moved not only beyond thinking about universal 'woman' to perspectives on 'difference', but also developed the concern to see ways in which women can find common bases for action and collaboration while honouring their diversity in experiences and contexts.

In the plenary speeches during the NGO Forum in Beijing (compiled and edited by Eva Friedlander) new issues emerged:

identity and difference; human rights; responsibility and accountability of governments, non-governmental and international organizations; and questions of institutional transformation . . . The speeches reflect the ways in which identities are played off one another to create multiple levels of similarity and difference, unity and divisiveness. As differentiation and inequality bring benefits to some while marginalising others, the question of how to create unity while accommodating difference becomes a central problem for the future of the women's movement.

(Friedlander, 1996)

The event in Beijing showed, again, how each participant (having come from a specific spot on the globe) was connecting her local web of social relations to the wider ones, and by doing so enabled the latter to connect with those in her place.

A parallel with the development idea

There is a parallel to be drawn here with the history of the idea of 'development' and its practice. For decades 'development' was defined in universal terms and thought of as a blueprint for the well-being of humankind. The 'developed' sat gloriously at the top of the Rostowian ladder, which was to be climbed by all human beings in order to attain 'the good life'. 'Development' was supposed to be desirable and applicable anywhere, at any time, for anyone. Unfortunately, however, this heroic ascension did considerable damage to nature, cultures and people en route.[13]

The universalist, and therefore reductionist, approach in development thinking and practice ignored the historical and cultural diversity of the various local environments targeted as its 'beneficiaries'. The many failures of large-scale development programmes, however, have slowly awakened a certain awareness of the need for a more culture-sensitive approach to problem resolution.[14]

The United Nations Decade for Cultural Development (1988–1997) which (significantly) followed the UN Decade for Women and Development and coincided with the UN Conference in Rio de Janeiro on Environment and Development, bears witness to the inevitable recognition that 'development' remains a myth as long as it is not tailored to fit *Our Creative Diversity*, the title given by the World Commission on Culture and Development, chaired by Pérez de Cuellar, to its final report.

Many development projects for women were initiated from outside by governmental or non-governmental agencies with a view to improving women's socio-economic situation. Experiences with these projects have made it more and more evident that women in different parts of the world – belonging to different cultural contexts, social classes and religions – may perceive their womanhood and their life (partly) differently, may act on that perception (partly) differently, and may even resist 'integration' into a development model which does not respond to their own perception of their own aspirations. Key issues in such resistance are so-called 'development' programmes which violate natural resources. One example of an environmental movement initiated by a woman is the Green Belt Movement in Kenya.[15]

The trend towards further differentiation can be seen as characteristic of a new phase. Initially, the 'women's issue' (like the 'development issue') tended to be stated in universal terms: it was thought to be the only way to make the point sufficiently powerfully. Now that the general analysis has to be applied to specific micro-levels, it is the diversity of situations which becomes more apparent, and – perhaps even more importantly – the differences in perceptions and appreciations of these situations. It is an attempt to explore this latter phenomenon and its implications that has led to the production of this book. This exploration has been conducted in a spirit of trying to understand rather than to explain.

The book project: purpose and expectations

The central idea behind the book project was a wish to highlight what women personally consider and appreciate as basic dimensions of their own lives, their sources of strength and the things that bring meaning to their lives. Rather than objectively addressing the socio-economic position of women in various cultural contexts, the book is concerned with women's own – subjective – perceptions of their environment and the forces which drive them in shaping their lives the way they do.

These environments are not only linked to the physical places where women are born and those in which they subsequently live. They are also made up of historical and cultural settings which in turn are shaped by complex patterns of changing relationships. Wherever people are located, their 'locus' consists of shifting connections with their direct environment and the world beyond it. The question of how women face, negotiate and shape the social space of their environment clearly needed to be looked into from a time perspective as well as a cross-cultural place angle. In other words: what changes do we see over a certain period of time in a given geographical context (multigenerational or 'vertical' perspective); and how does a particular period in which a generation experiences a certain environment make a difference (intragenerational or 'horizontal' perspective)? How does moving between places influence a woman's life; and how do changes in the place where she remains, perhaps all her life, affect her?

These questions were at the centre of the idea to combine the place and time dimensions by trying to find women from a variety of backgrounds (geographical, cultural, religious, socio-economic and age) who would agree to write about the lives of various generations of women in (or close to) their family.

The expectation was that these narratives would provide rich material for enhancing the understanding of similarities and differences in women's lives over time and across space. For instance, the grandmothers' generation was thought to have been more bound to its place of origin than the present-day generation of the authors and their daughters. It was expected that increased mobility might affect a sense of belonging, of rootedness.

Finally, it was hoped that what grandmothers, mothers and daughters in different parts of the world were going to say about their ways of shaping their

6

own realities would not only bring out differences, but also make explicit the underlying driving forces which many women have always implicitly recognized in each other and which hopefully will continue to create a common base for shaping a more humane society – at home and elsewhere.

THE PROCESS: A POLYLOGUE

The choice of the authors

The composition of the group was based on the criterion of achieving diversity in geographical, socio-economic, cultural, religious and professional background, as well as age. The names of the candidates were suggested by the members of an intercultural Core Group which had been formed to monitor the process. At this point, information on the candidates was restricted to their activities outside the family circle: next to nothing was known about their personal background. The women who accepted to participate turned out to be not only quite diverse indeed, as was intended, but also to share a level of education which allowed them to take up the challenge of writing the life narratives, to communicate in English or Spanish and to travel outside their country. Most (but not all) of them had followed some kind of higher education and are professional women now, even if their mothers were not.

It hardly needs to be stated that the group thus formed was not meant to be in any way representative of 'women in the world', nor of women in their own country, class, age or creed – not even of their family. Expectations did not go beyond the supposition that each of the invited participants would have her own specific frame of reference from which reality is observed and analysed, as well as her own personal approach to the common exercise. For instance, a Mexican politician, an Indian guru or an African rural woman would probably approach reality in a different way than a Finnish housewife, an Indigenous leader, a Muslim psychiatrist, or a European linguist. Not only might her perception be different, but also her conclusions. The combination of a variety of cultural, religious and professional angles might lead to a more comprehensive light being shed on 'the ways' of the women in the stories.

The fifteen authors were invited to meet for five days in order to share their perceptions of the life narratives. This Encounter was marked by a striking acceptance of the diversity among them. Their ways of writing and talking showed clear shades of difference between what has been called 'print-oriented' and 'oral/aural'-oriented modes of expression. There was a quiet willingness to listen seriously. Before the event, each of the authors had read the life narratives of the others: already familiar with these shared narratives, they had a strong sense that the mental distance between each other had already been significantly reduced before the geographical distance was removed and they finally met. It was then not a matter of identifying with each other because all were women;

rather, there was a feeling of being familiar with the unfamiliar, of being curious about the known.

Very little attempt was made to discuss in the original sense of *dis-cuatere*, 'to clear away by breaking up' or 'to examine and pass upon judiciary'. Nor was there any urge to convince others of the validity of one's own position. Instead, the interaction between the women could best be characterized by the meaning of the word 'dialogue': *dia-legein*, to converse, to immerse oneself mentally in someone else's situation and allow a questioning of one's own position.

This mental disposition shaped the Encounter in an unconventional and ingenious way. Presentations emerged which had not been agreed on beforehand, and what was agreed in advance could be changed in a very free-flowing and productive way. A topic was deepened if the conversation so demanded, and issues that had been put on the agenda but which did not naturally emerge were simply left out. All of this made for a vivid and profound polylogue.

The choice of life narratives

In view of the motivation behind the project, it was felt that the working method underlying the polylogue should reflect the need to stress differences in perceptions and appreciations of realities. In addition to this, the interrelations which exist between these different perceptions and appreciations, and the ways in which women shape their realities in different places and at different times, were considered relevant and important. These two concerns gave rise to the choice of the life narrative as a 'tool' for making explicit these differences and interrelations.

Women recording their own life histories, or those of other women, in order to understand more about their perceptions of realities is by no means new. The history of novel writing shows abundant examples of women writing about themselves or about other women and their representations of life.[16] The feminist movement has also stimulated an important renewed interest in oral history as a subject for research by, about and for women.[17]

The life narrative is the translation of a perception of events. It is intuitively analytical. It presents causes and effects in a selective manner. It refers to facts, experiences and accounts of facts, and reveals the sense which these make to the narrator. It shows how narrators perceive and shape their own and others' identities, and how they are thus guided in their actions.[18]

The narratives in this book mainly reveal the perceptions of the authors and, to a lesser extent, those of the women they write about. Some of the authors expressed more hesitation than others in terms of the moral dimension of selection and reinterpretation of events. Each found her own solution to this problem. Some discussed the text they wrote with the women concerned and/or with other family members. Some carefully selected the events they described and left out personal interpretations of them. Some gave only a minimum of personal information and insisted on the historical and socio-economic context. This diversity has been intentionally maintained, as it was felt to fit the objective of

8

leaving a maximum amount of space for differences in perception of what has been significant in the lives of the women described.

However, these limitations with regard to the tension between narrative and historical truth do not interfere with the purpose of this book. In the course of subjecting highly personal and selective life narratives to an intercultural exchange of thoughts in an interactive process, some of the implicit motivations behind the great variety of choices made by the women in the stories became explicit. These discoveries in turn revealed, unintentionally, some further commonalities.

The intent reader will discover for herself/himself nuances, subtleties, and the richness of differences in content, in events, in the ways of describing, in emphases, in questions raised and solutions found. So did the authors: their discoveries gave rise to different people being struck by different things, to currents and undercurrents sensed and identified. These were shared during the Encounter, and it was decided to share them with the readers of this book as well.

The life narratives differ in terms of length, style and degree of personal information. There are distinct individual emphases, diverse experiences and different perceptions of them. Although some language editing has been done, no attempt has been made to mould the narratives into a standardized style and format. This choice was made consciously so as to remain true to the observation of one of the participants: 'We met as multi-faceted individuals and not as uni-dimensional resource persons.'

The processing of the narratives

The first round: the personal in a historical context

To start off the process, the participants were asked to write a first paper describing briefly – preferably on the basis of interviews – the lives of their grandmother, their mother, themselves and their daughters (or, in case this was not possible, the lives of other women of these four generations close to the family). This description was to focus in particular on:

- those facts and factors which were felt to have had a determining influence on the lives of the women concerned
- what were seen as the major opportunities and constraints affecting their feeling of 'personal integrity' (of 'wholeness', a feeling of being complete)
- reactions to these opportunities and constraints.

This first round resulted in fifteen narratives grouping the stories of fifty-four women. They are situated in twenty-four countries and cover a period of over a hundred years. The Core Group responsible for monitoring the process came together and analysed the results of this first round. On the basis of this analysis a new set of questions was formulated for the second round. All papers of the first round were sent to all the participants with the new questions.

The second round: beyond the personal

All participants were asked to read the other fourteen narratives and to send comments or questions for clarification to the authors concerned. This correspondence proved to be a moment of discovery regarding the subjectivity of one's own perceptions: what had seemed self-evident to the author was not always so obvious to the reader.

In addition, the participants were asked to write a second paper. While building on the life stories, the authors were now asked to move from the descriptive to the analytical, from the personal to the collective, and make an effort to deepen the historical and the cultural aspects of their narratives.

Questions suggested for reflection in the second round of papers were:

• Given what you have written about yourself and others, what do you consider to be the major choices which women of your time and place have to face? What are the guiding principles, the values, implicit in these choices?
• It is often said that most societies are dominated by masculine values and that it is time for feminine approaches to have more impact on the way societies function. What would this mean in your specific context? What exactly should change?

The Encounter

The Encounter itself (held in Brussels, 12–16 October 1994) was envisaged as a distinctive stage and a further step in the thinking process. The products of the preparatory writings were not used as discussion papers, but as reference points.

The participants themselves set the agenda, by suggesting the issues to be discussed during the Encounter. The agenda was focused but open – focused in the sense that it aimed at finding some tentative answers to the main concerns behind the initiative of the book project, but open as to the choice of analytical frameworks used to help grasp the realities being described. This allowed for additional inputs from participants with regard to their experiences with women's ways of shaping their realities, and their analytical perspectives on women's problems in society. The result was the emergence of specific issues such as 'the female body as target for religious and cultural identity', the participants' own attitude towards spirituality and institutionalized religion, and the pitfalls of the use of dominant languages. These issues were explored during the Encounter itself.

By way of follow-up, the participants decided to continue the process after the workshop by allowing time for improving their written contributions which then could be published together in a book.

The book

The book itself consists of three main parts: this introduction, the life narratives and the findings. The life narratives are grouped regionally: they take the reader on a 'voyage' from Southeast Asia through the Middle East to Africa, Latin America and Europe. The start of the voyage in India is arbitrary. The third Part ('The Findings') constitutes an attempt to highlight some of the currents and undercurrents of the life narratives. This presentation should be understood as a wish to share with the reader the richness of the insights that emerged from the polylogue. At the same time, it is hoped that readers will discover for themselves other currents and undercurrents.

NOTES

1 Teresa de Lauretis has brought together the work of feminist scholars in the fields of history, science, literary writing, social criticism and theory, with the relation of feminist politics to critical studies as its general and overarching concern (see de Lauretis, 1986). She notes that 'there are a general uncertainty and, among feminists, serious differences as to what the specific concerns, values and methods of feminist critical work are, or ought to be'. In her introduction, she acknowledges that

> these debates make us uncomfortable because they give incontrovertible evidence that sisterhood is powerful but difficult, and not achieved; that feminism itself, the most original of what we can call 'our own cultural creations', is not a secure or stable ground but a highly permeable terrain infiltrated by subterranean waterways that cause it to shift under our feet and sometimes to turn into a swamp.

> (de Lauretis, 1986: 7)

2 'Women's rights are human rights!' exclaimed the First Lady of the United States at the 1995 United Nations Conference on Women in Beijing. Apparently, this still has to be repeated in spite of decades of worldwide attention to the lack of equal rights for women in many domains of public and private life. The main slogan of the *UNDP Human Development Report 1995,* which proclaimed that 'Human Development, if not engendered, is endangered', is beginning to sound equally repetitive. Add to this the fact that the World Commission for Culture and Development, in its final report, *Our Creative Diversity,* recommends that a concrete timetable be set for countries that have not yet signed or ratified without reservations the Convention to Eliminate All Forms of Discrimination Against Women (CEDAN). It may sound strange, but also strangely familiar: politicians' ways of shaping their realities . . .

3 The distinction between 'emancipatory politics' and 'life politics' is explained by Giddens:

> Life politics does not primarily concern the conditions which liberate us in order to make choices: it is a politics *of* choice. While emancipatory politics is a politics of life chances, life politics is a politics of lifestyle . . . In exploring the idea that the 'personal is political', the student movement, but more particularly the women's movement, pioneered this aspect of life politics.

> (Giddens, 1991: 214–215)

4 Gaining equality for women in the development process was high on the agenda of the Women In Development (WID) school from 1975 onwards. The second WID approach shifted the emphasis to redistribution with growth and basic needs. The third and most popular WID approach is to ensure that development is more efficient and effective for women (Moser, 1989).

It was not the concept of 'development' as such but the way it was put into practice which was put under the microscope by the WID school. Helen Brown criticizes what she calls the 'Liberal Feminist (WID) School' (beginning with Esther Boserup) as being biased by a 'modernisation approach to development, which originated in the West . . . , its optimistic unilinear evolution, the dichotomous conceptualisation of change, the diffusionist view of development and the reference to the West as the normative model of modernity'. She takes the position that

> the shift to 'efficiency' approaches within the WID framework marks the transition to a more overt instrumentality towards women on the part of the development bureaucracy, intensifying the subordination of the poorest women in Third World Societies. The liberal feminist WID policy derivatives are indeed part of the problem rather than a solution.
>
> (Brown, 1992)

5 Although the term 'gender' was launched in the early 1970s (Ann Oakley), it gained popularity only towards the end of the 1980s. It recalls Simone de Beauvoir's famous statement, in *The Second Sex*, that 'One is not born woman, one becomes woman' (*Le Deuxième Sexe*, 1949, Gallimard). 'Gender' is a theoretical construct, an analytical tool, designed to elaborate a distinction between the biological dimension of sexual differences and the cultural dimension of the division of tasks between women and men. As a theoretical tool the concept has proved to be useful for clarifying various roles (reproductive, economic, social), for defining differences in practical and strategic needs between women and men, and for applying these distinctions in a typology of approaches in development projects (Jacquet, 1995).

6 Chandra Talpade Mohanty is one of the many Southern feminists who criticize Western feminism in quite unequivocal terms. Her well-known article, *Under Western Eyes*, bears witness to a deeply felt frustration about 'the effects of various textual strategies used by writers which codify Others as non-Western and hence themselves as (implicitly) Western'. She suggests that many feminist writings

> colonize the material and historical heterogeneities of the lives of women in the third world, thereby producing/re-presenting a composite, singular 'third world woman' – an image which appears arbitrarily constructed, but nevertheless carries with it the authorising signature of Western humanist discourse . . . Sisterhood cannot be assumed on the basis of gender; it must be forged in concrete historical and political practice and analysis.
>
> (Mohanty *et al.*, 1991: 52, 58)

Southern feminists not only refuse to allow their cultural practices to be generalized and portrayed as 'feudal residues', and themselves as 'politically immature women who need to be versed and schooled in the ethos of Western feminism' (see Amos and Parmar, 1984). They also observe that Western feminism is obsessively preoccupied with sexuality.

This preoccupation led to considerable friction at the Women's Mid-Decade Meeting in Copenhagen in 1980 where Western women raised the issue of clitoridectomy and infibulation of female genitalia as a violation of human rights. Women from the Arab and African countries concerned took expressions like 'savage customs' as suggesting that their cultures are 'backward' and need to be Westernized. They also saw in this attack an implicit alliance with the rise of anti-Islamic political tendencies in the West. As a result women who were not themselves in favour of these practices felt that they were pushed into a position of defending them.

Cheryl Johnson-Odim formulates the problem thus:

> While it may be legitimately argued that there is more than one school of thought on feminism among First World feminists – who are not, after all, monolithic – there is still, among Third World women, a widely accepted perception that the feminism emerging from white, middle-class Western women narrowly confines itself to a struggle against gender discrimination . . . many have defined it as a liberal, bourgeois, or reformist feminism, and criticize it because of its narrow conception of feminist terrain as an almost singularly antisexist struggle.
>
> (Johnson-Odim, 1992: 315)

7 The Irish sociologist Ethel Crowley acknowledges the 'inadequacies of Western Feminism' for Third World women. She criticizes both Marxist and radical feminism as 'culture blind'. While stating that 'male domination is indeed universal', she believes that 'the void in feminism as we know it, results from the inadequate integration of the cultural dimension in political and economic analyses'. She adheres to the idea that 'culture is "the filter" through which we perceive the world around us. The social structure both creates and is created by the meanings attached to everyday aspects of life.' She then pleads for a reconceptualization of 'militancy' which acknowledges different female strategies in different contexts:

> Women often resist capitalist and patriarchal domination in anonymous invisible ways which may ultimately serve their interests much more efficiently than an overt challenge to the existing system.
>
> (Crowley, 1991)

A strong plea for specificity and plurality has come from female philosophers active in French intellectual circles, whose names are closely related to alternative ways of thinking and writing, called *écriture féminine*, 'feminine writing'. (This includes, among others: Hélène Cixous, Jeanne Hyvrard, Annie Leclerc, Marguerite Duras, Monique Wittig, Chantal Chawaf, Julia Kristeva and Luce Irigaray.) This movement, which goes back to the Parisian 'May revolution' of 1968, builds on insights developed by Lacan, Derrida and Foucault. Apart from their sympathy for Derrida's strategy of 'deconstruction', the *écriture féminine* is also characterized by a plurality of styles and a conscious effort to create space for the plurality of meanings. Reacting to the pretension of scientific language of being unequivocal and unilinear in its search for cause and effect, they emphasize the metaphorical components of language. *LA féminin*, the female feminine, focuses on the specificity of female sexuality, woman's authentic voicing of *jouissance*, 'pleasure', an embodied female subjectivity. The singularity of each woman is emphasized (see Irigaray, 1983).

As Denise de Costa lucidly explains in her book about Irigaray, Kristeva and Lyotard (de Costa, 1989: 46), 'within the "écriture féminine" movement it is thought that "LA féminin" can be found in places of oppression within the symbolic order: the non-logical way of thinking or the ambiguity of meanings; the unconscious which is culturally determined; the differences or multiplicity instead of unity and order; the bodily, in particular the female sexuality, etc.' (translated by Edith Sizoo). Although in both *écriture féminine* and postmodernist writing silences are speaking, one could say that 'the *écriture féminine* speaks where postmodernism keeps silent'.

8 See Braidotti (1991), Chapter 8, 'Radical Philosophies of Sexual Difference, or: I Think Therefore She Is'.

9 Many Southern feminists see the political aspects of feminism much more broadly than in relation to gender alone. Marie-Angélique Savanne, President of the

Association of African Women Organized for Research and Development (AAWORD), wrote:

> In the Third World, women's demands have been explicitly political, with work, education and health as major issues *per se* and not so linked to their specific impact on women. In addition, women in the Third World perceive imperialism as the main enemy on their continents and especially of women . . .

> (Savanne, 1982)

10 A useful analysis of the difficulty of counterbalancing the phallocentric nature of Freudian psychoanalysis is provided (among others) by Jessica Benjamin. Exploring the central question – Does woman have a desire of her own, one that is distinct in its form or content from that of man? – she comes to the conclusion that

> woman's desire can be found not through the current emphasis on *freedom from:* as autonomy or separation from a powerful other, guaranteed by identification with an opposing power. Rather, we are seeking a relationship to desire in the *freedom to:* freedom to be both with and distinct from the other . . . The phallus as emblem of desire has represented the one-sided individuality of subject meeting object, a complementarity that idealises one side and devalues the other. The discovery of our own desire will proceed through the mode of thought that can suspend and reconcile such opposition, the dimension of recognition between self and other.

> (Benjamin, 1988)

11 References to manhood, says Elisabeth Badinter, are most frequently expressed in the imperative mode. That familiar command 'Be a man!' implies that being (or becoming) a man is not as natural as it may seem. Virility is not given, it has to be constructed. See Badinter (1992) for a convincing description of how it may be much more difficult to be(come) a man than to be a woman.

12 Carol Gilligan describes the questions guiding her research as

> questions about our perceptions of reality and truth . . . about voice and relationship . . . I reframe women's psychological development as centering on a struggle for connection rather than speaking about women in the way that psychologists have spoken about women – as having a problem in achieving separation . . . I have attempted to move the discussion of differences away from relativism to relationship, to see difference as a marker of the human condition rather than as a problem to be solved.

> (Gilligan, 1993: xiii, xv, xviii)

13 Durre S. Ahmed analyses the phenomenon of the Western obsession with 'development' from a Jungian perspective as the 'hero archetype'. She argues that, although the story of the Hero is a universal one, it has taken a particular form when moving up from the Mediterranean to the north of Germany. Archetypally, she says, it involves the birth of a boy in unusual circumstances, separation from his origins at a tender age, the facing of tremendous trials and dangers as an early proof of super-human powers, then a return to his origins as a victor, a ruler, a unifier, a redeemer and a giver of laws. After some time, either through betrayal or hubris or heroic self-sacrifice, he falls from favour to decline and death. In psychological terms the myth is a symbolic, as well as a physical, enactment of the emergence of willpower and reason – that is, of rational consciousness. Durre refers to Jung and Hillman, who

suggest that these qualities were further enhanced by the co-optation and fusion with another powerful archetype, that of monotheistic religion. The latter is also based on, and moves towards, a principle of unity: its concern is individual morality, perfection and transcendence toward an ideal transpersonal future. This kind of moral reductionism and the fusion of the two archetypes provide the justification for all types of social action, including violence, against whatever seems 'outside' a prescribed idea of unity.

> The cultural locus of the ascension of a heroic and monotheistic consciousness over a fundamentally multiple and diverse personality can be said to be a movement from South to North. As Hillman observed, 'South' is both an ethnic, cultural, geographical place and a symbolic one. It is the Mediterranean culture with its images and textual sources, its sensuality and myths, its tragic and picaresque genres, and its stress on the feminine and cyclical nature of life. But the relentless upward march of the hero gradually overshadowed this fecundity, moving increasingly towards the style of Northern Epic heroism.
>
> (Ahmed, 1992)

However, since 'the heroic ego is blind to its own internal forms of violence' (Ahmed, 1992), it can cause massive damage on its way. This combination of upward surge and resulting wreckage has resulted in the coining by some Southern women of the expression 'e-wrec-tionism'.

14 The South North Network Cultures and Development brings together people working in base communities and researchers in Asia, Latin America, Africa and Europe who are concerned with the need for a culture-conscious approach to development processes. This Network publishes a journal (*Cultures and Development – Quid Pro Quo*), stimulates action-oriented research, provides training and offers consultancy. For further reading, see Verhelst (1989), Sizoo (1993), Rist and Sabelli (1986) and Sachs (1990).

15 The Green Belt Movement in Kenya was initiated by a woman (Wangari Maathai) and has now spread to thirty-five other African countries. In this movement women have planted Indigenous trees and set up community tree nurseries, thus providing much-needed fuel wood. See J.H. Momsen (1991).

16 During the last few decades, recording 'women's words' in songs, popular sayings and stories has received growing attention as a way of better understanding how women express themselves, and what they want to say about their own feelings with regard to their realities.

One among the many examples that could be cited is the tremendous effort made by Hema Rairkar and Guy Poitevin in collecting thousands of *cantelenes*, invented by village women in India thousands of years ago. These were sung at 4 a.m. or 5 a.m., when the women sat down, two by two, face to face, at the millstone to grind the flour for the meal of the day, while their men and children were still asleep. These songs represent a rare authentic and autonomous cultural production of women of olden times. With these songs they conveyed to each other their emotions and their views on life and womanhood. As the millstone has been increasingly replaced by mechanical mills, the songs are being sung less and less frequently; but they are now being revived through the work of Hema and Guy, and used in women's groups. See Kamble and Kamble (1994).

17 Feminist scholars of different disciplinary backgrounds bear witness to the efforts to develop a methodology for recording oral life narratives which is true to their principles of empowering women. See, for example, Berger *et al.* (1991).

18 Path-breaking work on qualitative methods of investigation for the study of lives and

life histories can be found in a series of publications edited by Josselson and Lieblich (1993, 1994, 1995).

REFERENCES

Ahmed, D.S. (1992) 'The Myth of the Hero', paper prepared for UNU/WIDER Conference on Systems of Knowledge as Systems of Domination. Republished in Durre S. Ahmed (1994) *Masculinity, Rationality and Religion: A Feminist Perspective*, Lahore: ASR Publications, Appendix IA, pp. 119–139.

Amos, V. and Parmar, P. (1984) 'Challenging Imperial Feminism', *Feminist Review* 17: 3–19.

Badinter, E. (1992) *XY de l'identité masculine*, Paris: Editions Odile Jacob.

Benjamin, J. (1988) 'A Desire of One's Own: Psychoanalytic Feminism and Intersubjective Space', in Teresa de Lauretis (ed.) *Feminist Studies/Critical Studies*, London: Macmillan, pp. 78–98.

Berger, G., Patai, S. and D. (eds) (1991) *Women's Words: The Feminist Practice of Oral History*, London: Routledge.

Braidotti, R. (1991) *Patterns of Dissonance*, trans. Elisabeth Guild, Cambridge: Polity Press.

Brown, H. (1992) 'Feminism and Development Theory: A Critical Overview', Occasional Paper 7, Department of Sociology, University College Cork, Ireland.

Butler, J. and Scott, J.W. (eds) (1992) *Feminists Theorize the Political*, London/New York: Routledge, Chapman and Hall, Inc.

de Costa, D. (1989) *Sprekende Stiltes: Een Postmoderne lezing van het vrouwelijk schrift*, Kampen, The Netherlands: Kok Agora.

Crowley, E. (1991) 'Third World Women and the Inadequacies of Western Feminism', *Trocaire Development Review*, pp. 43–56.

Friedlander, E. (ed.) (1996) *Look at the World Through Women's Eyes: Plenary Speeches from the NGO Forum, Beijing 1995*, New York: Library of Congress, catalogue number 96–67821, pp. xxvi.

Giddens, A. (1991) 'Self Identity and Modernity', in *The Emergence of Life Politics*, Cambridge: Polity Press, Chapter 7.

Gilligan, C. (1993) *In a Different Voice*, Cambridge, MA/London: Harvard University Press.

Irigaray, L. (1983) *Ce sexe qui n'en est pas un*, Paris: Editions de Minuit.

Jacquet, I. (1995) *Développement au masculin/féminin*, Paris: Editions l'Harmattan, pp. 31–60.

Johnson-Odim, C. (1991) 'Common Themes, Different Contexts', in C.T. Mohanty, A. Russo and L. Torro (eds) *Third World Women and the Politics of Feminism*, Bloomington: Indiana University Press, pp. 314–327.

Josselson, R. and Lieblich, A. (1993) *Vol. 1: The Narrative Study of Lives*, London/New Delhi: Sage Publications.

——(1994) *Vol. 2: Exploring Identity and Gender*, London/New Delhi: Sage Publications.

——(1995) *Vol. 3: Interpreting Experience*, London/New Delhi: Sage Publications.

Kamble, S. and Kamble, B. (1994) *Parole de femme intouchable*, preface Guy Poitevin, Paris: Editions Côté Femmes/Fondation pour le Progrès de l'Homme.

de Lauretis, T. (ed.) (1986) *Feminist Studies/Critical Studies*, London: Macmillan.

Mohanty, C.T., Russo, A. and Torro, L. (1991) *Third World Women and the Politics of Feminism*, Bloomington: Indiana University Press.

Momsen, J.H. (1991) *Women and Development in the Third World*, London: Routledge.

Moser, C. (1989) 'Gender Planning in the Third World: Meeting Practical and Strategic Gender Needs', *World Development* 17(11): 1799–1825.

Rist, G. and Sabelli, F. (eds) (1986) *Il était une fois le développement*, Lausanne: Editions d'en Bas.

Sachs, W. (1990) 'The Archeology of the Development Idea', *Interculture* 109, Montréal: IIM.

Savanne, M.-A. (1982) 'Another Development with Women', *Development Dialogue* 1(2): 8–16.

Sizoo, E. (1993) 'L'Aide et le pouvoir', *Economie et Humanisme* 325: 28–25.

Verhelst, T.G. (1989) *No Life Without Roots*, London: Zed Books.

Part II

THE LIFE NARRATIVES

2

ETCHINGS ON A GRAIN OF RICE

Kamala Ganesh, India

Janaki, 1906–1969

Padmavathi, 1921–

Kamala, 1952–

Gowri, 1965–

JANAKI, 1906–1969

Diffusing boundaries between home and world

In her later years, Janaki 'mata' (mother) as she was known, had established a well-known ashram (a sort of commune focused on spiritual activity) in Thanjavur, Tamilnadu. She was a guru to many men and women who, years after her death, continue to revere her memory and come together on special days to celebrate and commemorate at a temple which has been consecrated over her *samadhi*, where her body was interred at Thanjavur. Janaki mata considered herself to be in the spiritual lineage of Ramana Maharishi, whose philosophical enquiries into the nature of the self and whose own life at Tiruvannamalai made him a highly respected sage with a multitude of followers from India and abroad.

At the age of 13, Janaki was married to a widower twenty years her senior, who had two young daughters. He was a respected government doctor and was, I am told, a loving husband. They shared a lifelong relationship of mutual regard. Theirs was considered to be a 'modern' and 'forward' family. Janaki spent some years in Vienna with her husband; back at home, she was a badminton player, active in the ladies club, and so forth. Running parallel to this, Janaki seems to have undergone a series of strange and mystical experiences, right from childhood, and to have displayed a capacity to perform miracles; this convinced her and others that she was a special person with divine qualities, meant for some great task.

Many of these experiences involved physiological and psychological disturbances and traumas, and phases of trance-like absorption.[1] The impression I have

22

is that her journey to spiritual self-awareness was very disruptive of normal life. Even within a fairly religious family environment, the tumult within her for spiritual expression was not easily comprehensible to others. Duty-bound as wife and mother, tied to this-worldly activity, she did not know how to give expression to her yearnings. She left home in search of a guru, and met Ramana Maharishi, who was able to assuage her anguish. He persuaded her to return home, assuring her that spiritual growth was not incompatible with family life. She settled down in Thanjavur, running an open house, which eventually acquired the character of an ashram. Men and women engaged in similar quests – as well as orphans and those in need of refuge – found a place there. Some were itinerant, some permanent settlers. Her personal charisma, and bold articulation of her views through talks and writings, made her a popular and sought-after guru. Notes and diaries kept by devotees talk about how people would bring all manner of personal problems to her. She demanded that they put their total faith in her, and surrender. As if by a miracle, the problem would be solved. I have also heard of instances when she would empathetically experience a devotee's pain, transferring it on to herself, leaving her/him cured and free.

Members of her own family also lived with her. Along with a prescribed routine of meditation, prayer, ritual, worship, poor-feeding and charity, there were also celebrations of festivals, family occasions, illnesses and pregnancies, quarrels and conflicts, children going to school and college, and so forth. Her children married into families with 'normal' backgrounds, and several of her grandchildren grew up in the ashram. She was close to them, as she was to her devotees. When her husband died in 1955, Janaki went into deep introspection. She left home, ostensibly to drown herself in the sacred water of the Ganges river, but was persuaded to come back and continue to guide her disciples and devotees. In her later years, she had a troubled relationship with her eldest son, who grew up to be a rationalist and atheist. He saw his mother's donning the mantle of 'guru' as fake and egocentric. He was also angry with the way she had overturned domestic hierarchy, at what he saw as the marginalization of his father. Janaki stuck to her beliefs; they were not reconciled in her lifetime. Her youngest son developed mental problems and died young under strange and unexplained circumstances.

Janaki lived for many years in Thanjavur, deified and worshipped by her children and devotees. For the former, she was mother but also guru. For the latter, she was guru but also mother. The best-known gurus today have an international network of disciples, travel extensively and mobilize huge funds for their charitable and social work projects. Often, with multimedia promotion of thoughts and words, the marketing hype seems to dwarf the substance, even for the highly respected gurus. With Janaki, the time period and the local, small scale of her functioning enable one to grasp better what elements of 'tradition' and 'religion' came into play in allowing her what was certainly a substantial amount of leeway to carve out a niche for herself.

To me, the central contradiction of Janaki's life lies in her efforts to fuse her

23

role as mother (who creates and nurtures a home) and as guru (for whom the entire world is home). In a patrilineal and community-centred ethos, mother-hood carries a high load of sacrifice and care-giving, to be channelled exclusively into a specific family and home. On the other hand renunciation, in its widest sense, seeks to break the bonds of personal relations defined by kinship and gender. The ideal renouncer should therefore never have been married and should cut ties from the family of birth. The ritual of performing funeral obse-quies for one's own self dramatizes this. Many orders do accept those who have been married and had children, but they have to sever these ties at the time of renunciation.

Janaki did not go through the rituals of formal renunciation. Such a route, through any of the prescribed sects and orders, is not normally available to women.[2] Janaki's own guru Ramana had not taken this path either; he could be seen more as part of the tradition of 'saints' rather than *sanyasis*. Nevertheless, Janaki lived her life in ways that would have been unthinkable, not just unaccept-able, for a woman in her time and space, had these not been lodged within the overarching framework of religion. She was able to reach out to the public sphere and also plumb deep into her own inner self. Many women – not just her contemporaries, but mine as well – would find themselves boxed within a zone that does not facilitate either. Given that she did not have any 'backing' from established institutions, she achieved her status by her own luminosity, presence and actions, for which the broad culture did give her scope. She was a public figure and had independent relationships with men and women outside the kinship network. She was the focal point of their lives, commanding loyalty and devotion as a leader. She was the centre of gravity of the household, and over-turned the essence of the hierarchies of deference due to husband, sons-in-law and other senior males, without compromising on external courtesies.

But Janaki was also a mother and a grandmother in the full sense. In this she was different from the celebrated women saints of earlier centuries, like Mira or Mahadeviakka, who had renounced home and cut bonds of family in what constituted a more straightforward interpretation of self-realization. Janaki chose instead to link the public and domestic. Dissolving boundaries between home and world, yet maintaining a distinction, and interweaving detachment with attachment, called for a feat of intellect and emotion: a sleight of hand through which one's mental and spiritual poise was deemed uncircumscribable by phys-ical and material circumstances. Her last child was born after her first grandchild, when she had already become a spiritual initiate of Ramana. Her grandsons recall playing on her lap, even when she was at the peak of public life.

More pertinently to my argument, she also chose the idiom of 'motherhood' to relate to her disciples. Culturally, this was perhaps inevitable, but it seems also to have been a great personal need. She was no reclusive ascetic, concentrating solely on individual liberation through spiritual austerities. She took great plea-sure in cooking and supervising the kitchen and personally serving food to the ashram members and visitors. She would oversee the deliveries of her daughters' and granddaughters' children. She would get closely involved in nursing the sick. This is quite different from our image of the male guru, who would (typically) distance himself from physical intimacies and focus on mental techniques of self-control. Janaki's spirituality was expressed through an affirmation of the immediacies of life, and an active involvement rather than a withdrawal. The root of her feeling of personal integrity was her orientation towards nurture and care, extending the boundaries of motherhood to encompass a larger commu-nity, outside the biological. And yet her gender did not result in a devaluation of her sphere of action. It is worth considering that perhaps there did exist, in her time, a public sphere not shaped by male values alone. But it hinged on the use of the idiom of motherhood.

Janaki's mode of action simultaneously articulated another contradiction. On the one side was spiritual pursuit, with its stress on individual effort for salvation, and on freedom from social categories – a lineage traceable to the medieval bhakti movement. On the other side were religiosity, established dogma and hier-archy. The course of the lives of her followers, tending towards the latter, suggests that her energies were eventually absorbed within the framework of brahmanical Hinduism which is hierarchical and patriarchal in its structure and assumptions. In India, while there is a profusion of expressions of religiosity on the part of women, reinforcement of domestic roles is often the unspoken condi-tion on which this is predicated. In Janaki's own life, and in her actions, as against the interpretation of these by her followers, the balance was tilted away from social categories. The dexterity with which she welded the contradictions, and the conviction which she carried, argue for more than individual heroism. They suggest the existence of structural openings which she harnessed. They militate against a single or simple definition of Hinduism.

PADMAVATHI, 1921–

Like water on a lotus leaf

As the eldest daughter of Janaki, and as a loyal disciple, Padmavathi has striven to live according to her mother's philosophy. Janaki's tremendous influence on her never ceases to amaze me: the degree of courage and serenity which she brings to the surface in moments of crisis have surely been drawn from this source. This is especially impressive, because by disposition Padmavathi is (or was) a timid person, self-effacing, something of a worrier, constantly seeking support, happy to lean and share. With advancing years, she has developed considerable autonomy in the conduct of her life, though she presents a mild exterior. Part of this meek persona was perhaps picked up in the course of playing various familial roles. Padmavathi has mentioned that as a young bride with a strict mother-in-law and a strong-willed husband who was professionally very successful, her best option (her only option) was to avoid expressing firm opinions and try to win over her new relatives by acquiescing with them.

Throughout a good part of her married life, she played a supportive role to her husband's professional and familial responsibilities. Her husband was a source of financial and other kinds of support to his (and, in time, to her) relatives, and there would always be a nephew or cousin staying at or visiting their home. The ambience of a joint family makes it difficult for the wife/mother to pursue any intense activities other than housekeeping. But Padmavathi had her own agenda, too: she was keen to visit Thanjavur frequently, for which she required her husband's permission. Her almost unnatural attachment to her mother, and her penchant for a certain kind of religious expression, must have been looked upon with suspicion by her mother-in-law as much too self-indulgent for a young wife. Janaki must have been perceived as a dangerous force pulling her away from her duties to her new family. I do not know what kind of subtle communication went on among the four protagonists, but over the years Padmavathi did manage to pursue her interests, keeping her bonds with her mother intact. She had won over her husband. Three of their children grew up and studied under the care of Janaki.

Padmavathi lost her husband when she was 46. Today, with grown-up children and grandchildren, and with five great-grandchildren, she leads an active and independent life. In contrast, the majority of her contemporaries are far more dependent on their sons for physical and financial security. (Unusually for her times, Padmavathi's husband left her financially secure and independent.) She spends part of the year at Janaki's ashram, immersed in its activities. She makes brief visits to the homes of her children in different parts of the country, travelling on her own. She likes to visit places of pilgrimage and can manage very well alone in strange places, provided they have some significance for her. She is keen on performing acts of charity and piety. She reads a great deal on topics connected with the Hindu religion and philosophy, and constantly brings what she has read into conversation, with a level of articulation unusual for a person with only a school education.

Padmavathi has addressed the dilemma of being a wife, mother and grandmother while simultaneously pursuing a path of self-realization predicated on the principle of detachment from worldly entanglements, including money and children. She follows the solution that Janaki developed – not giving up one for the other, but integrating them. This quest has informed Padmavathi's life and has been the source of her personal integrity. Janaki's quest had greater originality and conviction: it shook up her life like a cyclone. She had a bold personality that could transform the dynamics of a situation and become its hub. Padmavathi was no aspirant for guruhood: she achieved her ends more indirectly, without disturbing existing arrangements. When she is in a mood to articulate the models that women should emulate, she draws upon accepted imagery of patience, sacrifice and putting family first. She stresses the importance of subtle and skilful human relations. Women must make their way not through confrontation, but indirectly and gently – advice that often comes from women in patrilineal and community-oriented contexts. Her own life trajectory and some of her choices have been, in practice, 'individualistic'. Neither she, nor those close to her, perceive it this way, however, because these are embedded within an overarching spiritual orientation.

For Padmavathi too, the detachment/attachment mix ('like water on a lotus leaf, on it but not in it' – a frequent allusion in philosophic expositions on how to achieve *moksha*, or 'liberation') expresses itself through the 'care' idiom: cooking, feeding, serving, nursing in illness, counselling in difficulty. But it is not to be contained within any one family for too long. While Padmavathi gets involved in her children's families at times of crisis (like illnesses) or special occasions (like weddings), at other times she avoids becoming entangled in domesticity, and keeps her own time for meditation, reading, discussion, listening to religious discourses, and so on. I know her only in this phase, but imagine that when she was in the middle of raising her own children, she did not detach herself so much. In other words, while her current interests are abstract, contemplative and inwardly-directed, her intuitive responses to the immediacies of a situation are concrete, active and practical, and she draws upon the traditionally feminine modes of nurturance.

Over the years, Padmavathi has broadened her attitude towards 'modernity', reducing her involvement in dogmatic aspects of religion. In matters of faith, she tends to be attracted towards the magico-sacred realm; for her, this coexists with an acceptance of science and technology in practical matters. 'Tradition' and 'modernity' are thus cast in complementary terms, a legacy that has coloured the approach of subsequent generations within the family.

KAMALA, 1952–

In search of personhood

When I got married, the whole ethos described above came as a shock; the more so since there was first subtle and later direct pressure on me to conform to and reproduce it. Although my own natal home was also 'traditional', it was not so explicitly religious, and the onus of bearing the load of tradition was on my mother. My father, a railway official known for his high personal and professional integrity, had a thirst for knowledge and learning and this influenced the atmosphere at home. We girls were encouraged to excel in our studies, to read, to develop talents in languages, music and painting. But, in contrast to the boys, there was no serious career planning. For us, education was meant to broaden horizons, and to be convertible to income only as a contingency. We grew up in a closely supervised and protected environment. Because of my father's frequent transfers, my siblings and I were exposed to different regional cultures, although these remained mostly middle-class, urban situations.

At home, a fairly strong orientation to Tamil culture prevailed. My mother's formal education had ended with school finals, and she married at 15. She had a fairly active social life connected to my father's professional circles, but the bulk of her time was taken up with housework, bringing up the children and taking care of my father's mother, his widowed aunt and her invalid daughter, who would spend extended periods with us. Mother was also a keen observer of festivals and ceremonies. In contrast, my father – at least in his youth – was not very ritualistic although he was well-versed in philosophical texts. He has, of late, developed deeper links with certain aspects of 'orthodoxy' which are compatible with a rationalistic world-view. But neither of these things – obligations to kin or the call of faith – impinged on the children's routine in any very obvious way. Discipline was strict but matter-of-fact; we fell in line with whatever was required of us. I cannot remember feeling deprived or constrained, even though in retrospect I feel that our lives were too ordered. Father was strict but fair, and mother worked hard and ran the house systematically. In different ways, both were constantly trying to enhance our skills and talents to ensure future betterment.

I was a bright student, but not very interactive, always aspiring to be a good, model child. I think I apprehended events in an intellectual, distanced way. I must have been reacting to the influence of teachers, peers and others, but I do not recollect any sharp upheavals until my late teens, when I started reading about 'women's lib' in *Time* magazine and the like, and started connecting it to real-life situations. I did not want to get married and live on the terms and conditions that my mother seemed to be tied to. I wanted to study further. I was slowly developing an interest in social sciences, even though I had graduated in Chemistry. Around this time, my parents were looking seriously for a marriage alliance for me: when they sensed my fear of the expectations attached to a traditional 'arranged' marriage, they flew into a panic and redoubled their efforts to get me married before I became too stubborn. They argued that if I let opportunities slip by when I was young and eligible, it would become progressively more difficult to find a suitable match. I was confused, but it did not occur to me that this was a negotiable issue and that I could oppose it. Ingrained in my mind was the idea that living in my parents' home was conditional upon following their wishes, and that if I refused, I should leave and fend for myself – which seemed unthinkable then. My feeling of resentment against my father lingered for many years. My younger cousins and nieces have been more vociferous, although 'arranged marriages', under somewhat liberal conditions, are still very popular in our family.

While I had reservations about marriage itself, I was impressed by the man I married, by his personality and the fact that he was so articulate. However, this was also my period of intense reading of feminist literature and I was critically analysing every person and every situation that I encountered from a feminist perspective. Sparks flew in our marriage. I was opposed to having a child, since I felt that this would constrain my freedom to study and work, and in any case I had problems with the idea of giving birth in a patriarchal situation. When I became pregnant soon after marriage, I unilaterally decided to abort; this caused a rift between us for many months. Although kind and gentle, my husband's mother was insistent on some expression of religiosity on my part. I reacted negatively to her efforts at proselytization. While I recognized her moral claim on her son, I felt that I could not be treated as his extension. Her approach to him on the subject was roundabout: I would try to counter this through direct confrontation with feminist arguments. My husband's response was vague. He would talk about the need to please elders, about how no bond could be built without sacrifice. He would say that one cannot bring textbook arguments into real life and convert it into a college debating arena. I was surprised that a US-educated engineer with radical political leanings could say this.

During those years, I was burning with a desire to find an interesting and meaningful job which would also make me financially independent, but was confused as to how to go about it, considering the non-professional nature of my qualifications. My husband, who was very involved in his career, seemed indifferent, and my mother-in-law was polite and indirect, but hostile. I had a small child by then, and no firm arrangements for child care. I had, in the last few years, fumbled my way through a few research projects, and acquired a postgraduate degree in Sociology. When I read about a tiny community in the southern tip of India where the women lived in strict seclusion, I jumped into registering for a doctorate. In my frame of mind at the time, the subject presented issues that instantly appealed. I left my small son with my mother, and spent several months in the field and several years on the project.

The experience of fieldwork was something new to me and, together with my involvement in some feminist organizations in Bombay, led me to rethink several assumptions I had made earlier about the situation of women in India. I became more alert to context and nuance. My perceptions now are that gender subordination is interwoven with other kinds of inequalities and privileges, and needs careful unravelling. People's own ways of apprehending and dealing with their situations cannot be easily categorized, but are criss-crossed with ambivalence, compensations and hope.

My professional interests since then (projects, writing, teaching at university) have been at the interface of anthropology and women's studies. While my encounters with academics and activists from different countries (through my work at the Commission on Women, International Union of Anthropological and Ethnological Studies) have helped me become sensitive to 'other cultures', this has, paradoxically, made me more introverted, more enmeshed in my own context – familial, regional and national. I feel convinced now that one has to be in active engagement with one's immediate realities, even when one is trying to handle conceptual and analytical modes that transcend the specific. I believe that there are strands in the history and culture of the subcontinent that provide inspiration as well as workable alternatives for women seeking to change their lives. The 'Chipko' movement against deforestation in the Himalayas, and SEWA (Self-Employed Women's Association) in Ahmedabad are the better-known examples of women drawing on Indigenous resources to transform their reality, but there are so many flashes and fragments that one keeps encountering individual women in everyday life. Janaki's life, and Padmavathi's too, affords glimpses of this quality.

My own personal inclinations in the last few years have been in the direction of greater involvement in intimate ties. I now have two sons, and motherhood has opened up a whole new dimension of experience. Sharply aware of, and concurring with, the feminist critique of mothering in patriarchal systems, I nevertheless feel humanized and rehydrated by becoming a mother. Perhaps under the cumulative weight of all those formidable ancestresses, I have given my children, especially in their early years, preference over professional advancement whenever it comes to the crunch (which it does with amazing regularity). Functioning in a largely nuclear set-up, it seems to me at such moments that I have a chance of making some difference to my children's lives. A report or paper that I am writing, on the other hand, has a remote chance of making any difference to anybody. Under the circumstances in which academia functions, it is more than anything else a way of self-expression. In the meantime, our stormy marriage has become more companionable. Initially, this was due to my husband's patience, even though he felt antipathy towards feminism. At the moment, we are both concerned with providing a stable environment for our children and share a commitment towards building on, rather than erasing, what we share. He has, I notice, changed some of his perceptions on women and feminism, without actually saying so. In another sense, so have I. It has come as a discovery to me that within the parameters of a conventional marriage one can create areas of freedom.

GOWRI, 1965–

Adding value to the symbolic capital of 'family'

What I only found after lengthy and painful exploration, Gowri seems to have assumed from the very start. The course of her life so far suggests that she has avoided casting ideas in opposition to each other: tradition and modernity, religion and feminism, marriage and freedom.

Gowri has had a privileged upbringing as the daughter of a celebrated and respected sportsman who is something of a legend in the country. She is a qualified lawyer, has played competitive tennis, trained intensively in the classical dance of Bharatanatyam and travelled, from childhood onwards, to many countries. Her father's values of focused striving for excellence have no doubt shaped her personality. Her brother's upbringing was dominated by a single-minded pursuit of excellence in sports: in her case, the emphasis was more on adaptability, resourcefulness and developing skills in relating to people. The source of this differential emphasis is most likely her mother, who has been influenced both by her grandmother, Janaki, and her own experiences as a young bride in a family which, from ordinary beginnings, achieved great success and visibility through relentless efforts in its chosen field. The singular focus on the part of the males was underpinned by its absence among the females, who provided the support system for the high achievements.

The ambience of Gowri's girlhood was infused with a considerable amount of religious activity (a mixture of social, ritual and spiritual) through her mother, running parallel with education, travel, friends and other ingredients of a modern lifestyle. Gowri's mother was different from her own mother (Padmavathi) in that, although religious, she downplayed the 'personal salvation' component and lodged herself deliberately in the domestic arena. Unlike water on a lotus leaf, she has concentrated entirely on the nitty-gritty of building a home, moulding a champion, nurturing relationships of which she is the linchpin, and adding value to the symbolic and social capital of the family. She sees a woman's position not as one of weakness. The domestic sphere is one of power, achievable for women through commitment and sacrifice, and her sense of personal integrity is enmeshed in its fortunes. Power in the public realm pales in comparison with domestic power; it is bereft of meaning without it and certainly does not supersede it. Domestic power is, in other words, a territory that no woman should let go lightly. Gowri too has taken on this understanding, although at the moment it remains incipient.

In her early twenties, Gowri went in for an arranged marriage to a career diplomat. I was initially surprised that with all the opportunities available to her she had not taken the initiative to find her own partner. But I realized that she had reflected carefully on the matter, and had participated actively in her parents' efforts. She had not accepted the first proposals that came, but had waited and discussed the issues with her parents. Having spent long periods in countries of the West, she had decided that the popular Indian system was better for her. Within these parameters she sifted the possibilities, and chose what struck her as workable. Earlier generations of girls would not have dared to express such definite views.

In recent years, Gowri has been travelling to different countries, bringing up her two small children, and coping with the social demands that accrue to a foreign service wife. She has been chafing at her inability to pursue anything seriously – a career or an interest – given that there is high peer pressure in her generation to be something other than 'just a housewife'. She has been trying to get back to tennis and Bharatanatyam.

Gowri has had her share of 'adjustment' problems with her husband and in-laws, stemming mainly, I think, from the contrast between her liberal upbringing and their more traditional expectations. As far as I can judge, Gowri has chosen the course of investing her energies into her family, maintaining kith and kin networks and creating a supportive atmosphere for her husband's career, while at the same time not fully accepting their definitions of what she should be. She has been able to actualize a high level of choice and achievement along with a tacit – sometimes verbalized – endorsement of 'Indian' values. She is actively engaged in creating a space for herself in her marriage. Its precise contents will surely be different from those of preceding generations, but she retains the primacy of her

claim to the domestic realm in ways both like and unlike her mother, grandmother and great-grandmother.

REFLECTIONS

Is there a larger story?

In extrapolating from the specific to the general, the question must inevitably
arise: Which women is one speaking on behalf of?

I have chosen to tell the stories of women not related to me by blood. My
memories of my own grandmothers are hazy; I have no daughter, nor do my
siblings. I have known my husband's mother for the last twenty-two years. I know
about her mother, through the 'tellings' of several close relatives and disciples,
and through her own writings, for she had a larger-than-life persona. I have
watched the daughter of my husband's sister grow, from being a child, into
adulthood. For an unbroken matrilineal chain, I should also have included her
mother (my sister-in-law); but somewhere in my rendering of their lives I felt a
need to insert myself. Perhaps this would light up some part of my own motivations in producing such specific versions of their stories, which they would surely
have told in other ways than mine.

In its intimate and intricate detail, each is a very individual story. Yet we share
a certain milieu, by birth and by marriage within the same community, arranged
by elders. Upper-caste Tamil-speaking families of our kind were among the
earliest to take to English education in the British period, and are known for our
emphasis on high educational achievements. Technical and professional education for women, however, was not encouraged until recently. Occupationally, the
stress was on professional and government service, with a marked reluctance for
entrepreneurship. For married women, careers were virtually ruled out until ten
to fifteen years ago. By two generations ago, such families had moved from a
rural to an urban base through all-India migrations (and currently count among
the more prominent groups of migrant Indians to North America). While this
made for increased exposure and a certain 'cosmopolitan' outlook, southern
India was also historically sheltered from many of the major social and political
upheavals experienced by northern India. It was, in cultural terms, relatively
more conservation- and stability-oriented, until the advent of early twentieth-
century radical social and political movements, notably the Dravida Kazhakam
and its derivatives. Brahmin families such as ours, who had historical access to
the 'high culture' of the region, stressed values of learning, knowledge and
philosophy rather than social reform and political action.

Our families were socially conservative, and the domestic ethos was characterized by clear conceptual separation between men's and women's spheres.
Cooking and feeding, nurturing and caring, domestic religious rituals and an
extraordinary sense of continuity and stability constitute the images of 'home'
for me, derived not just from my own home; 'extraordinary' because, with

hindsight, one knows that these families were not immune to upheavals from within and without. A sense of continuity was created through management of change and conflict, and in this the investment of women was high. Some common features characterize the objective familial environment in all four stories, and these are by no means commonplace across class and strata. There is a fair amount of financial security and tolerance for the individual's particular quirks and inclinations. None of these women, for instance, has faced an actual situation of ' . . . or else you can leave the house', a threat which is not impossible, given patrilineal rules of space. Within this class, there is considerable social pressure, internalized by women, to maintain a semblance of family harmony and sexual propriety.

In all four stories, there is a creative tension between religious/spiritual and secular spheres, and between a lodging in collectivity versus individualized pursuits. In each set the first is apparently overarching, but the dynamics are different for each person. For three of them, the choices are not yet complete, but continue to shift in the space between the poles. It is tempting to construct an oasis of domestic stability, a charmed lineage of continuity, that somehow endures despite the fact that one is talking of people who lived in different times and situations. But 'eternal India' is only a semi-myth. I have focused on those partial truths, which to me are the most interesting aspects of the stories: the expression of connectedness to the small, the everyday, the micro-world, and an intuitive recognition of its vitality. This is set against the backdrop of a massively variegated country and civilization, and tumultuous social, political and economic changes in the last century, in which women's lives and perceptions too have been shifting and changing. The four women are obviously from different generations, not to be conflated, nor to be seen as 'representative'.

I am conscious that I live in a society and culture where traditional hierarchies have been superimposed with new ones. In a somewhat contrived exercise, I create an *alter ego* out of Vandana, and ask: Does this reflect her reality? My sources of knowledge (intimate, albeit fragmentary) about women from poor strata are the domestic helps who have worked at my house in Bombay at different times. Vandana in some ways typifies them. Widowed three years ago (her husband – unemployed and an alcoholic – died of jaundice), she has two young children who are looked after by relatives in the village. She stays in a slum not far from where I live, and works as part-time domestic help in seven or eight houses, from 7.30 a.m. to about 6 p.m., usually with a short break in between. She earns perhaps a quarter of what I earn from my university job. Vandana is completely unlettered, has tattoos on her arm and chews tobacco. She also uses English words and phrases like 'problem', 'blood pressure', 'tension' and 'My God!' in her conversations, with a precision of meaning. In her dark visage and slim, tough body, my agile imagination sees a link with her bronzed forebear from Harappa, hand on hip, unfathomable smile on 5000-year-old lips. She certainly belongs to the urban proletariat, with highly unprotected, insecure conditions of work, living in a squalid settlement on the margins of a megalopolis. I like to think that beneath the routine drudgery of sweeping, cleaning and washing in eight houses she responds to life crises and arrives at decisions from her elemental store of protohistorical memory, not unspliced with current street savvy and survival sense.

But, between us, Vandana and I do not capture anything close to the tremendous diversities in India, which create qualitatively different choice structures for women. The economic divide, the rural/urban/tribal divide, are generally congruent with a sharp educational divide. Cross-cutting traditional hierarchies of caste, regional culture is a powerful unifier. Religions, languages, and kinship and marriage systems bring in further cleavages, where earlier they were more diversities than cleavages. Vandana's four-generation story, were she to write one, would be no more true of women in India than mine. Given the irresolvable problem of representation, an attempt to pick out general issues relevant to women's concerns will have to be a self-conscious one. I therefore propose to discuss briefly the issues of 'working women', the role of religion in women's lives, and the need for valorizing 'female values' into non-gendered components of public life.

My contemporaries and I have had the opportunity of a good education, although not necessarily one which would equip us for a professional career. But to choose whether to 'work' or be 'just a housewife' is a dilemma. Many have chosen something light or part time, which can be combined with family and children. In this social segment, there is a great deal of investment in the education and upbringing of children as routes to future social mobility, further sharpening the dilemma. Although a mixture of kin, community, State and market resources are drawn upon for child care and housework, these options are not well-structured ones, unless a woman has professional qualifications and a well-paid job. The responsibility for care-giving for children and elderly parents is taken quite seriously – in contrast to lateral relatives, towards whom responsibilities have shrunk. Across strata, women do think of themselves as embedded in kinship, and their sense of personhood is connected to fulfilling these obligations, even though the detailed contents of the obligations have changed. The career vs. home dilemma thus remains a source of some anxiety, although (increasingly) younger women in this segment are qualifying themselves for high-powered careers. This particular dilemma appears to be resolving itself, with 'working women' becoming an accepted category.

At another point in the spectrum, women like Vandana have no choice but to earn for the sheer survival of their family. Given the low level of skills and training, the frequent absence of a male breadwinner, and a precarious balance between income-generating work and household responsibilities, it is difficult to exercise options regarding the type of job. Women from this group are often deeply committed to educating their children as a way, however tenuous, out of their present privations – although how far this extends to girl children is uncertain. Across strata, the principle implicit in the eventual decision regarding women's work, especially in the case of married women, is that of enhancing the family's resources as a whole (including its social and symbolic capital), rather than individual gratification.

In part, the life of Janaki mata and her influence on subsequent generations accounts for the focus of the life narratives presented here. Yet, even in the larger context of feminism, the linkages of gender with identity politics and the issues of religion and culture are ones that bear serious consideration. Much of the writing on women and religion in India has asked the question: How does religion view women? For the most part, the answer has been that (ideologically, normatively, legally) women have an inferior position compared to men, with some compensations thrown in (see, for example, Mukherjee, 1978). This has received reinforcement with the recent and vigorous scrutiny of, and debate on, personal laws based on religion (for a sample, see Kishwar, 1994; Agnes, 1994). The undeniably negative scriptural injunctions have also induced a school of scholarship which posits an earlier golden age (especially with regard to Hinduism) in which women's status was high, followed by decay and deterioration (see Altekar, 1962, among others). That the (religious) ideology conceptualizes women not in a singular way, but as multiple and contrasting evaluations, is discussed less (for a sensitive exception, see Allen, 1982); yet even this, like more unequivocally critical evaluations (such as Chakravarti, 1993), remains within the parameters of scriptural and normative sources. As Falk and Gross (1989: xv) point out, in this genre we are never told anything about women as human beings, the texture of their lives, and whether such views correspond to their own self-concepts. Some recent writing on women and religion in India has come out of the reductionist bind of looking at normative positions alone, and has focused attention on the ways women experience religion in their lives – for example, its possibilities as a source of meaning and empowerment (Eck and Jain, 1986; Ganesh, 1990); as an agent of transformation (Dietrich, 1986); and for expanding operative space for women (Falk and Gross, 1989).

Contemporary issues of religious and cultural identity are intertwined with the politics and history of the subcontinent. For women under various kinds of subordinations and restraints, the exercise of identity (even when some of its political links are problematic, and its cultural content selective) seems to open possibilities for positive self-affirmation. In the alienating and dehumanizing life conditions of women like Vandana, but also for others (men and women), it offers scope for a sense of anchorage and renewal. An agenda for culture-sensitive feminist action in the Indian context should, in my view, be concerned with retrieving life-affirming aspects of religious traditions from their automatic clubbing with obscurantist moorings or the more sinister political appropriations.

With all awareness of context-specificity, one can still glimpse certain qualities among women that cut across strata: what one could – tentatively and carefully – call 'female values', even though they are neither biological nor exclusively female nor necessarily possessed by all women. And yet, by sheer weight of historical association, they are overwhelmingly female. The quality of involvement in immediacies and in detail, in the small and insignificant, comes through entanglement in life processes which will brook no distancing or abstraction. The absolute need to reproduce micro-world activity every day leads to the value of relationing and harmonizing situations and space, and to a circular and qualitative concept of growth and progress. The vital charge that these values carry is not unrecognized, but it is relegated to private, domestic and gendered spaces from where it is loaded with the onerous task of cushioning the impact of the 'real world' in which the macro-values of scale, competition and success are valorized. My personal interpretation of feminism leans towards the degenderizing and deprivatizing of 'female values', and their entry into the public sphere and the macro-world, rather than the mere entry of women into this sphere as honorary males on the terms of the macro-world as they exist today.

NOTES

1 I have not gone into this aspect in detail, partly because my own argument here is confined to examining how 'religion' and 'spirituality' were limiting and simultaneously liberating to Janaki's location as wife and mother. Partly, I was uneasy about getting flooded by devotees' recounting Janaki mata's miracles, an aspect which I did not want to claim kinship with. But, on reading Yogananda's autobiography (1994), as well as the lives of Ramakrishna, Anandamayee ma and so on, I was struck by the similarities in the phases of spiritual development and by how much 'madness' and 'abnormal' behaviour are part of this; also how, in spite of scientific and rational explanations by the guru, the basic premiss of her/his teachings provides a terrain for the acceptance of miracles and suprabodily phenomena. One of Janaki's disciples has written about how *he* reacted to her presence: at times, for no reason, tears would flow uncontrollably; for days he would suffer pangs of hunger, no matter what he ate, and then for days, he would go without food; he would go into *bhava samadhi* for long spells, during which he would be oblivious to everything around him, moving about as if in a dream, or would withdraw into deep meditation. He would go through periods of physical agony when his entire body seemed to be on fire (Gowri, 1990).

2 Sex is considered as an obstacle to reaching spiritual goals (except in tantra). In the framework of renunciation, women are inherently evil, tempting men away from the straight path. They are also permanently in worldly bondage, through their enmeshment in bodily processes of reproduction, and cannot take formal vows of *sanyasa* (normatively). Guru Jnananda and gurumayi Chidvilasananda are exceptions in current times. The former was reportedly allowed to take *sanyasa* vows by the last Shankaracharya of Kanchipuram, considered to be one of the most orthodox and authentic centres of *sanyas* (White, 1989: 16). The latter was not only initiated into *sanyas* but also installed as the head of the Ganeshpuri centre by her predecessor Muktananda himself. (For an account of Muktananda, see White, 1974.) There is also a rich tradition of saints, both male and female, who do not belong to an established or formal order. Some modern monastic orders do permit women to join.

REFERENCES

Agnes, F. (1994) 'The Hidden Political Agenda Beneath the Rhetoric of Women's Rights and Uniform Civil Code', paper presented at a seminar on Nation, State and Identity: A Post Ayodhya Perspective, Majlis, Bombay.

Allen, M. (1982) 'Introduction: The Hindu View of Women', in M. Allen and S.N. Mukherjee (eds) *Women in India and Nepal*, Canberra: Australian National University Monographs on South Asia 8.

Altekar, A.S. (1962) *The Position of Women in Hindu Civilization from Prehistoric Times to the Present*, Delhi: Motilal Banarsidas.

Chakravarti, U. (1993) 'Conceptualising Brahmanical Patriarchy in Early India: Gender, Caste, Class and State', *Economic and Political Weekly*, 27(14): 579–585.

Dietrich, G. (1986) 'Women's Movement and Religion', *Economic and Political Weekly*, 26(4): 157–160.

Eck, D. and Jain, D. (eds) (1986) *Speaking of Faith: Cross-Cultural Perspectives on Women, Religion and Social Change*, New Delhi: Kali for Women.

Falk, N. and Gross, R.M. (eds) (1989) *Unspoken Worlds: Women's Religious Lives*, Belmont, CA: Wadsworth Publishing Co.

Ganesh, K. (1990) 'Mother Who is Not a Mother: In Search of the Great Indian Goddess', *Economic and Political Weekly*, 25(42–43): WS 58–64.

——(1993) 'Breaching the Wall of Difference: Fieldwork and a Personal Journey to Srivaikuntam', in D. Bell *et al., Gendered Fields: Women, Men and Ethnography*, London: Routledge, pp. 128–142.

Gowri, M. (1990) *Guru Mahimai*, Thanjavur: Sri Janaky Nilayam.

Kishwar, M. (1994) 'Codified Hindu Law: Myth and Reality', *Economic and Political Weekly*, 29(35): 2145–2167.

Mukherjee, P. (1978) *Hindu Women: Normative Models*, Calcutta: Orient Longman Limited.

White, C.S.J. (1974) 'Swami Muktananda and the Enlightenment Through Shakti Pat', *History of Religion*, 13(4).

——(1989) 'Mother Guru: Jnananda of Madras, India', in N. Falk and R.M. Gross (eds) *Unspoken Worlds: Women's Religious Lives*, Belmont, CA: Wadsworth Publishing Co, pp. 15–24.

Yogananda, P. (1994) *Autobiography of a Yogi*, Bombay: Jaico (first published 1946).

3

CHANGING FACES OF TRADITION

Durre Sameen Ahmed, Pakistan

Fatima, 1900–1962

Qamar Ata-Ullah, 1917–

Durre Sameen Ahmed, 1949–

Eman Ahmed, 1971–

FATIMA, 1900–1962

(as narrated by her daughter Qamar)

Her husband fell in love with a 'modern' woman and Fatima was given a divorce.

Defeated by tradition

Like most women of that time Fatima learned to read the Koran at home and was literate in her native language, Urdu. Along with four brothers and two sisters, she lived a middle-class life in the small town of Sialkot in what was then undivided India. At the age of 14 she went through an arranged marriage and moved to Kashmir. Her husband was the only son of middle-class parents and aspired to emulate the ruling class, which was, at that time, the British Colonial presence. Fatima found herself caught between two sets of demands: on the one hand, an archetypally powerful mother-in-law who regarded Fatima as little better than a slave and expected her to do all her chores, dutifully and silently; and on the other hand, a husband who was keen that she become 'modern', give up the veil and generally participate in a more glamorous lifestyle. But he was unable to stand up to his mother, and his job (as a forest officer) meant that he was away for long periods, leaving his wife at the mercy of his mother. Within a year they had a daughter, Qamar. By the time Fatima gave birth to her second child, a boy, about three years after the marriage, events were overtaking her. Her husband fell in love with a 'modern woman' who had come for a vacation to Kashmir. She smoked, danced, rode horses and they quickly decided to marry. Fatima was given a divorce and returned with the two children to her family in

40

Sialkot. Shortly afterwards, the boy died of pneumonia. Fatima spent the next thirty-five years with her parents, serving the household in such central ways as cooking, cleaning and sewing. When her own parents passed away, she came to live with me, her daughter Qamar; she stayed until her death, in her early sixties, from complications caused by diabetes.

Fatima's life can be seen, in the first place, as one of sheer powerlessness. She had no will of her own: girls were taught to be obedient, especially to their in-laws. But returning to her own family with two young children was no better. She was an excellent seamstress and could have managed to support herself and me, but this was not permissible, because divorce was equated with a kind of living death. Fatima was considered tainted and had to surrender to the dictates of family. She had no income of her own. Although she sewed clothes for the whole family, she was not allowed to do it for money: had this been possible, we could have been our own masters.

Similarly, she had no freedom in terms of raising me. The entire family felt, perhaps well-meaningly, that I was their responsibility: everyone tried to 'guide' me, usually with no regard for my feelings or those of my mother. But the family was the only security we had. For thirty-five years Fatima single-handedly served her parents and her siblings, as midwife, tailor, cook, cleaner. This was not seen as a contribution to the household, but as the household doing her (and me) a favour, because of her circumstances. Fatima had a tragic life.

QAMAR ATA-ULLAH, 1917–

The most important factor was my innate desire for learning and a steadily consolidated spiritual impulse.

Sustained through tradition

As a child I was keenly aware of my mother's situation and her life was a major determinant of my own. I realized she faced enormous constraints and limitations, and in trying to help her I developed habits – doing things myself even now, despite having domestic help – which go back to that time. Life in my grandparents' home was not easy and one was constantly aware that falling short of any standard – of conduct, studies, housework – would not be tolerated. The major debt I owe to my mother was her insistence that I should study, in which she was supported by my grandfather. So, while I helped in whatever way I could, I also went to school. I did well and, after a mild tussle, I was allowed to join a woman's college in a city 300 kilometres away. Before leaving for life in the college hostel, I promised my mother that I would not put pressure on her to provide me with nice clothes and so on; pressures which were bound to arise in an environment largely comprising very well-to-do young women. I was one of a tiny minority of Muslim women who obtained a BA in (still) undivided India. After the BA, I was keen to study medicine, or even teach. But my grandparents and mother said, 'Enough'; I should now gain a deeper knowledge of the Koran and prepare for marriage. I complied. Those two years were difficult: I was not really allowed to read or write what I wanted; letters which came from friends I had made in college were screened. I was confined, as if I were in a cage, and felt suffocated, helpless.

There were many proposals of marriage for me but I was not particularly keen on the idea, given my mother's experience. There was the added constraint that I should marry someone who belonged to the same Islamic sect as my family. Finally, the choice fell on a man who was twelve years older than myself. He was the friend of an uncle and had been married with two children; but the marriage had failed and he had been living alone for many years. While this was not in his favour, he was well-known and liked for his upright character. Also, he was a doctor with high status in the British Medical Service. Most importantly for me, he was planning to go abroad, to Europe, to specialize further. When the time came for me to give my formal consent, I agreed. In a sense, it was a matter-of-fact decision, based essentially on the assumption that I was exchanging one cage for another (but perhaps bigger) one.

The early years of marriage were difficult but also exhilarating. I travelled to Europe and my husband taught me about many 'wordly' things – from eating with a knife and fork to an appreciation of Titian and Beethoven. At the same time, he was highly orthodox and expected me to conform to all the traditional forms of behaviour, dress and demeanour. I have no shame in admitting that I felt bitter towards my in-laws, who would have preferred this marriage not to take place, but my husband made it clear that if I was to get along with him, I also had to get along with them. After the birth of our first son, I was on the verge of leaving (even my mother suggested it), but I had vowed that I would not follow that route, and somehow I found the strength to cope. The compensation was that my husband was a man of great intelligence and moral strength. He encouraged me to continue reading, which I have done voraciously all my life. We travelled widely on different assignments, although we also spent periods in what was for me the difficult and unpleasant milieu of my mother-in-law.

In 1947, India was divided. We lost almost everything we owned and came to Pakistan, where I had the last of my children – a daughter, after four sons. When she was 5 (and I was 35) I went back to formal studying, for an MA in English Literature. This was followed by an MA in Social Work and then another from the University of Chicago. I went to Chicago for a year on my own. Though my husband initially resisted this idea, he eventually agreed and, together with my mother, he looked after domestic affairs during my absence. On my return from Chicago I took up teaching at the university, where I remained until I retired at 60.

I was widowed fifteen years ago, but my life has continued to be blessed and full. All my earlier hardships have been compensated for and I enjoy a life of great comfort, security, and the blessings of my children and grandchildren. In retrospect, I could highlight three factors that had a determining influence on my life. The first was my mother's own life of extreme hardship and the discipline and determination I learned from her and my grandparents. The second factor was my husband, who taught me a great deal and encouraged me to remain a perpetual learner. The third factor, and possibly the most important, was my own apparently innate desire for learning and a spiritual impulse which has steadily consolidated. Forty years ago, as a student at the University of Chicago, I wrote (on 21 January 1953) the following as part of an assignment on 'my life':

> I am not worried for my children except for one thing. I am worried for the spiritual values of life, perhaps that is my only headache. The type of life we led was always based on strong spiritual values, which the present age and present day children are sadly lacking. I want to arrive at a solution where the pattern of life for my children would be based on love, confidence, security and also spiritual values. If this balance is achieved, then they would be able to look after themselves and I would worry no more.

I still feel the same and have tried my best to communicate these values to my children.

DURRE SAMEEN AHMED, 1949–

The earlier battles between my mother and myself have given way to an urgency to learn all that I can from her.

Defying tradition, regaining tradition

I am the youngest sibling and only sister with four brothers. Although my father was a profoundly religious person, he took great pleasure in introducing me as his 'fifth son' and encouraged me to do and learn whatever took my fancy. I do not remember feeling any constraints and had a delightful, rough-and-tumble childhood. I learned to ride horses, play the piano, shoot with a rifle, swim, and so on. In hindsight, there were many paradoxes in my life. I went to a convent run by French nuns who were appalled at my tomboyish behaviour. School years were one long attempt to discipline me and I survived the nuns only because of my other activities, my father's support and his diplomacy with the nuns. My compensation for the misery of school was an atmosphere at home which placed great emphasis on what was considered the best of Western culture: its music and literature, its thinkers, and a general reverence for science and knowledge. Although my parents also tried to inculcate in us a familiarity with Eastern culture, this was largely overshadowed by the external environments of school, media and a genuine respect for the West. Today, I realize the extent of this loss and am trying to fill in the gaps.

The only problem of that time was my mother and her growing alarm at the way I was 'turning out' as a teenager. There was a constant struggle between us as she tried to domesticate and feminize me, wanting me to dress 'properly' and not be so boyish. But with my father's help I prevailed, and grew up convinced that women were useless creatures until they had mastered the masculine world.

Around the age of 17 both my parents decided that I should get married. A cousin had proposed and was found acceptable by all standards. He had a degree from MIT, liked Western classical music (a condition which I had imposed, believing that no Pakistani could meet it) and belonged to the same, small Islamic sect as ourselves. This last factor was exceedingly important to my parents and, given everything else, he seemed ideal. Although I initially resisted the idea, I eventually agreed, for two main reasons. My mother and I had reached a high level of confrontation regarding my femininity, and I saw an opportunity to escape. This idea of escape was further enhanced by the fact that the man was studying and working in Holland and I saw the marriage as an opportunity to see Europe, the land and the culture which I inhabited mentally.

Naturally, no marriage could last on such a rationale and once I had 'seen' Europe I informed my family that I had had enough. Perhaps if my husband had been more amenable to my continuing education, it would have been different. This was an unhappy period of my life. At first, my parents were not willing to hear of divorce, so, after returning from Europe, I spent two years in an extended-family set-up. I read and read (perhaps too much!), all sorts of 'heavy' intellectual material; and eventually I had a breakdown. During this time I gave birth to my daughter. Finally, on the advice of a psychiatrist, my parents agreed to a divorce. I was 21 years old.

Although this must have been a trauma for my parents, I was never punished or made to feel guilty. I was encouraged simply to carry on with my studies and a host of other activities. I did my BA and, spurred by my own experience with therapy, followed this with an MA in Clinical Psychology. I then married a man who shared my love of language and literature; I started doing all sorts of 'womanly' things for my husband and wanted more children. For some reason, I wanted to be this way, to play this role – but discovered that my husband preferred a more masculine mate!

I was 25 years old and utterly confused. Despite my bravado, I knew that social expectations were high. Moreover, I wanted the marriage to work. So, for the next ten years the relationship continued, in theory, but I was constantly restless and unfulfilled. I went away for a year to study the performing arts in China; the separation helped and we had a son when I returned. But somehow the idea of marriage as a settled life, with many children, just did not take shape. As I write, I am aware that I do not want to blame my husband; obviously my own personality also played a part.

Partly out of necessity and partly out of frustration, I spent the next eight years acquiring a range of academic credentials (four MAs and a doctorate from Columbia University, New York) and trying to make the marriage work. It did not: at the age of 35 I returned to Pakistan to embark finally on a career (as a teacher/psychologist) and to end a long but unsuccessful relationship.

It was an extremely painful and difficult decision but I had changed in a number of crucial ways. From my teenage contempt for women, I had learned to appreciate and value them. Ironically, this happened more during my seven years in New York than as part of my own culture. It was the experience of living alone in a large, impersonal city with my daughter, and the effort of creating a home in difficult circumstances, that increased the nurturing side of my nature and made me more appreciative of women. In the process I began to enjoy my own unique experience of being a Pakistani woman in terms of dress, food, and so forth.

Today, I live with my mother and my two children in an extended family which includes the families of two brothers. The earlier battles between my mother and myself have given way to an urgency within me to learn all that I can from her, particularly about religion and spirituality. My various accomplishments are really hers, since she has supported me in so many ways. I am truly fortunate in having had the parents I did (my father passed away fifteen years ago). They were both important factors of my life. Another determining influence for me, still related to my parents, has been an insatiable appetite for knowledge and an optimism which is rooted in a deep and abiding faith in God.

EMAN AHMED, 1971–

I want to make something of myself before I devote myself to a husband and children.

Between tradition and modernity

The memories of my childhood are full of fun and games and the petty fights which children invariably indulge in. Having grown up in an extended family, I never lacked companions. Although I only have one younger brother, there was never a shortage of playmates. True, I may never have had the love and affection of a real father but my grandparents, my uncles and stepfather more than made up for it.

Up to the age of 7 I lived in Pakistan and led a very protected life. Then I went to live in America for three years with my mother, who was at that time a student at Columbia University. I do not remember much of my stay there, except that it was very different from my life in Pakistan. In Pakistan a woman was hired specially to look after me, but in the USA my mother worked and studied full time, so I had to take care of myself – sometimes even preparing my own meals (which usually meant climbing up on to the kitchen counter and taking down a tin of macaroni cheese!).

Compared to life back home, we did not live in great style; but despite living in Harlem, I attended one of the best schools in New York, Rudolf Steiner. Since most of my classmates came from very affluent families, they could afford luxuries I could not. Now that I look back, I believe those years contributed in a positive manner towards making me what I am today: I learned to be my own person, to stand up for myself and to speak out if I wanted to be heard – even scream sometimes! The experiences of growing up with seven cousins in Pakistan, and then in the USA, instilled in me the fighting spirit.

Around the age of 9 I came back to Pakistan and was sent to the same school my mother had attended, a convent. I felt that the nuns tried to suppress me. The school toned down my fiery temper and disciplined me quite extensively. Of course, I interpreted their efforts to discipline me as restricting me. College was fun. Having studied at an all-girls school, and given my family and Pakistani society's conservative outlook, going to a coeducational institution was a very exciting experience.

One important event in my early life was meeting my biological father when I was 14 years old. My family claims that it was a bad experience but, quite frankly, I do not remember much of the meeting. I believe that he agreed to meet me only after a lot of persuasion. My grandmother says that children tend to block out unpleasant memories, and perhaps she is right. I am often asked whether I miss not having a father and my answer to that is quite simple. I feel that you are only capable of missing things which you have once had. How can you miss something that was never yours to begin with?

Society is only one factor which has influenced my life; other factors include my home environment and the way I have been brought up. My family is generally liberal and progressive, but at the same time it is conservative enough to hold on to old values which have been handed down for generations. One of those values is that a woman's first priority is her home and family. However, the family is progressive enough to encourage its female members to study and be self-sufficient. My grandmother, mother and nearly all my aunts went to university after they got married. Having grown up in such surroundings, I was not very clear about the woman's role. This confusion was deepened further by the fact that I was raised by my grandmother during the early part of my life and then by my mother during my teens. My grandmother, due to her upbringing, believes that there are certain things which girls and women should not do: she does not approve, for example, of girls going out of the house after dark. In fact I feel rather sorry for her, because my mother has not fulfilled her ideals of what a woman should be. My mother is not really concerned with her appearance and I cannot remember the last time she wore jewellery. Another talent which a typical Pakistani woman is supposed to acquire is the art of small talk. That is one thing my mother just cannot do.

Having been raised by two quite diverse personalities I went through different ideas of what it means to be a woman. It is only now, working for a Women's Resource Centre, that I feel I am more balanced and clear. The nature of my work is such that it has forced me to sit down and think in a realistic manner about my role as a woman in our society. Entering such a field I have come into contact with many die-hard feminists who reject the notion that women belong in the home, that women are, above all else, daughters, wives, mothers, that there is no need for them to be self-sufficient since they are meant to be dependent upon their fathers and husbands.

Until a couple of years ago, getting married ranked high on my list of priorities. However, I now feel that I want to make something of myself before I devote myself to a husband and children. I want to establish myself as a person and not only as a woman. I want to be able to support myself if necessary. Once I am married, I will have no qualms about being dependent, emotionally or financially, on my husband.

Finally I would like to say that as a woman I have not experienced any discrimination by society. My experience of my uncles, etc., is that they feel protective towards us, and I do not think that the desire of our male relatives to protect us can be called 'discrimination'. But it does raise questions: Why do they want to do this? Do they feel that we are not equipped to protect ourselves?

I think it is a combination of many things. There is a desire to protect the values which they have inculcated in us. Basically, it is an old-fashioned family – not in the negative sense, because they do encourage female education and argument, and neither my mother nor my grandmother could be called oppressed – but old-fashioned in the sense that they don't want their values to be changed through the women of the family, so they protect their values by protecting us. Perhaps they also feel that I am not equipped to deal with the world as yet, so it becomes a vicious circle. Also the society we live in is such that if one person is 'pointed' at, in terms of respectability and honour, that finger is pointed towards the whole family. In that sense, they are also protecting themselves. But I do not think my male relatives do this deliberately or selfishly: there is a genuine sense of caring for the women.

REFLECTIONS

The question of choices and values is more of an issue for those of us born in the twentieth century, and these choices and values change in emphasis during individual development. One can only wonder about how my grandmother (Fatima) felt and thought about herself, but it is evident she did not have the range of choices that the subsequent generations did. It can be argued, of course, that the attitude of believing oneself to be helpless is a choice in itself, but I find such a view rather harsh and permeated by a dangerous 'Prometheanism'. Either way, it seems that Fatima's guiding principles were a desire for security linked with a sense of personal responsibility for her daughter and family. Significantly, her sense of duty towards her daughter was expressed in a fierce determination to educate her.

My mother, Qamar, seems to have made different choices at different times. The decision to marry a man she had never seen was made in order to 'exchange one cage for another (perhaps bigger) one', that is, for greater independence and a chance to travel. My own first marriage was based on roughly the same logic. Unlike me, however, once married Qamar chose to remain in what was an emotionally difficult situation, at least for the first decade. She was sustained in her hardship, and was eventually able to overcome it, by her deep faith and interest in spirituality and religion, along with an intense desire to continue a secular education. It was her practice and ever-increasing knowledge of religion that gave her inner strength and a sense of 'rightness' about life. Thus, while she spent the first ten to fifteen years of her marriage putting her own personal interests on hold, this was followed by a steady flowering of her abilities and interest in higher education and teaching. From being the shy, unsure young girl married to a much older and more experienced man, she developed her own achievements and a distinct identity. In her own assessment, this maturing and eventual balance between marriage, family and work was primarily due to her faith in God as expressed in the practice of Islam. What was equally significant for her was that this faith and practice found a friend and companion in my father.

I am a witness to this all-embracing spiritual dimension of the lives of my parents. To say that I was raised in an ambience permeated by a sense of the presence of God may evoke stereotypes about Islamic 'fatalism' and 'the Orient' – a picture of women with their heads covered, people in a constant state of prostration on prayer mats, the loud and incessant chanting of the Koran, the air heavy with incense. This was certainly not the case and yet, the cliché of Islam 'as a way of life' was very evident. While there was indeed a transmission and practice of ritual, this was basically complete before my teens; after that, it was largely left to individual choice. The only exception to this was the pre-dawn prayer, for which we were awakened and which was said together in my parents' room. None of us was ever forced to adhere to the other four prayers. There is an inherent ritual flexibility to Islam in which every point on earth is sacred and every person is her/his own priest. Thus, even as I do the same thing today, I recall my parents praying routinely in the midst of all types of activities – music, television, children. At the same time, I was also aware that they both spent long hours in prayer alone, late at night, a few hours before sunrise. Thus, while religion *per se* was not prominent, God was ever-present; not just as a belief system but as part of life, like eating and sleeping; a part of conversations on a day-to-day basis as a presence with which one interacted formally and informally. Dreams were among the topics which were regularly shared at mealtimes, and were given a spiritual interpretation.

At this point I would like to clarify certain commonly held perceptions about religion and spirituality, since I shall be discussing the guiding principles and values from within this context.

The Islamic faith

In the current historical circumstances, with an escalation of hostility between Islam and the West, it is important to note that the word 'Islam' has at least four meanings, depending on the context. The most important is the literal one – 'submission to God's will' – and its attendant connotations of 'peace through surrender'. The second meaning is in the context of the last and true divine religion, the final expression of Abrahamic monotheism based on the Torah of Moses, the Psalms of David, the Gospel of Jesus and the Revelations of the Koran. The third meaning is the one used by historians to describe Muslim civilization in the same way that 'Christendom' is distinct from 'Christianity'. Finally, its most common current usage is in the context of ideological conflict in many Muslim countries where 'Islamic' is used to imply 'the good', while its opposite 'un-Islamic' is frequently equated with notions of an overly permissive 'pro-Western' stance. In my view, this most recent usage of the term, Islam-as-socio-political ideology, dominates Western perceptions of Islam, while for the vast majority of Muslims, and certainly within the context of my family, it is the first, literal meaning which is central – that is, submission to God.

The guiding principles of Islam

To say that all religions ultimately talk of one Truth is correct, but as a substantive answer to an explosive and important subject it is not enough. It is like saying that all humans have two arms, two eyes, one nose, etc. Yet at close quarters each of us is distinct, in features, temperament and taste. Similarly, every culture and religion is indeed a constellation of numerous guiding principles and values, but they are different in what they emphasize.

I have found Fritjhof Schuon's work on comparative religion very useful in this connection, especially his analysis of the dominant psychological features in each of the monotheisms – even as they speak of the same God. According to Schuon, within the categories of, for example, fear and love of God, Judaism emphasizes the former and Christianity the latter. Again, it is a matter of emphasis, not an either/or dichotomy: both factors are present in these traditions, but one tends to be emphasized more. Similarly, for Schuon the emphasis in Islam is on knowledge. While this may sound rather odd given the state of education in the majority of Muslim societies, one should remember that this present gaping chasm between the essence of a religious teaching and its collective manifestation is a problem of all of the religions, and also of science. In fact, one could say that currently all three monotheisms are, in the social context, a complete inversion of their identifying essence. Instead of fear – which has to do with an idea of not exceeding limits – there is complete fearlessness; instead of love, there is hate; and instead of knowledge, ignorance.

Returning to the guiding principles of my mother and, to a lesser degree, of myself, I would say that it was the search for knowledge as inspired by Islam (and hence encouraged by my father) which motivated many of our choices and values. As I shall explain later, the awareness of these principles, for me at least, was not a conscious one until recently. Within this essentially unconscious Islamic *Weltanschauung*, there functioned a related principle which can be summed up by an Arabic attribute of God, *Al-Rahman*.

While most Westerners regard the Arabic 'Allah' as a sort of alien deity in the same category as Jupiter or Zeus, the fact is that there is simply no word in Arabic for 'God', and 'Allah' is the name of God in the same way that Yahweh is the God of the Old Testament. The attribute of *Rahman* is one among the numerous attributes of Allah and, because of the structure of the Arabic language, it is not possible to give a precise translation. It is variously translated as The Beneficent One, The Merciful One, The Gracious One, The Compassionate One. Its etymological roots suggest deeper meanings in its relationship to the womb and sanctuary.

It was this attribute of God, together with numerous other names-as-attributes, which permeated my consciousness. The smallest of events and situations were linked to Allah's grace and mercy. A Mozart symphony, a bath with hot running water, a meaningful dream, a luscious mango, and the capacity to be grateful for this – all were linked back to God's *Rahman*. Needless to say, especially in times of pain, loss and crisis, the same was evoked to assist in an attitude of forbearance and patience.

It is worth reiterating that, while describing this atmosphere/attitude of 'God-awareness', I do not want to paint a picture of pious self-righteousness regarding myself or my parents. My personal narrative is evidence that I have frequently fallen abysmally short of standards. Nor would I describe my parents as saints. The only way I can explain this is to repeat the notion of Islam as way of life, but one which is understood differently by different individuals.

My grandmother Fatima exemplifies the stereotype of Islamic 'fatalism': in other words, she seemed to accept passively the decree of divorce and the consequences following from it. It is difficult to assess to what extent her attitude was moulded by cultural constraints – a mixture of Hinduism and Islam – since divorce is not permitted by Hinduism but is allowed in Islam. Nevertheless, defeated as she was by the social constraints of her time, she was fiercely insistent that her daughter should receive the best possible education.

There is no doubt that the guiding principles in my mother Qamar's life were firmly grounded in Islam, but in a much broader and deeper manner than with Fatima. On the one hand, it was her faith that sustained her through the difficult early years of marriage, enabling her to tolerate hostile in-laws and an exceedingly demanding husband. But after the first ten years of marriage, she slowly began to pursue her lifelong passion for learning, culminating in three MAs and a career in university teaching. This dimension of asserting her identity also drew its strength from Islam and its emphasis on 'seeking knowledge, even if you have to go to China'. Thus, it was from within the Islamic perspective that she had the support of her husband, even to the extent of going off to study in the USA for a year. Similarly, one of the major events in her life, the death of her husband, was met with a grace and dignity the well-springs of which were primarily spiritual. Today she firmly believes that, despite her love for him, his death was her gain in so far as she has greater leisure time to devote to reading and spiritual practice, even while continuing to lead an otherwise active life.

Until my late thirties, I do not think I was guided by any one particular principle. Life was simply an adventure and there was a passion to learn and absorb a range of subjects and activities, which my father especially encouraged. While our overall lifestyle was simple, no expense was spared in satisfying my voracious appetite for books, music and travel. I don't think I gave too much thought to my first marriage, or having children. Looking back, I see only two consistent strands in an otherwise rather fast-paced existence, spanning change, continents and cultures. The first was an insatiable desire for learning, education, knowledge. The second was what my friends perceived as my 'eccentric' insistence on prayer, which I continued to do regularly, even when leading what could have been considered an impious life.

Thus, even though I was in fact living out the Islamic injunction as regards the primacy of knowledge and at least a ritual submission to God, I think that until my mid-thirties religion was a largely unconscious affair. Unlike my mother, who had studied sacred texts and writings from an early age, I was content to pray, fairly regularly, and leave it at that. God was simply a given, a friend, the Indulgent One (another attribute of Allah). I was encouraged, not necessarily to follow what were perceived as Western values, but rather to reach out to what was, logically, the best there was to learn. However unconscious these choices, the seeds of those principles pertaining to knowledge and spirituality were evident when one considers that even as a student of psychology I became deeply interested in the work of Jung.

As I approached my forties, however, I began to realize, primarily through my clinical practice, that many 'modern' ideas of knowledge were in headlong collision with the spiritual side of humans. It was a knowledge which asserted that 'God is dead' and which viewed human nature in extremes of black and white. And, while Jung provided an invaluable counterpoise, unfortunately he seemed to be rather ill-informed about Islam.

As these issues became increasingly apparent in my clinical practice, I grew more aware of the extent of this dichotomy as it existed not only in modern psychology, but in my social and intellectual relationships. It was only because of my mother, and her extensive knowledge of spirituality as related to indigenous Islamic conceptions of health and healing, that I managed to retain both my faith and the considerable intellectual discipline gained in Western universities. Then I began to realize more consciously that one guiding principle for me had been knowledge irrespective of cultural origin, and that it was time for me to apply that knowledge to what had, until then, been a largely unconscious relationship to my own religion. Thus I began to try to understand religion *per se* and Islam specifically at a more conscious, intellectual level, by undertaking a study of its history, psychology, aesthetics. It has been a most fulfilling adventure in learning, a true gift from the Gracious One.

Present/future choices

Parallel to these personal events, Pakistan has seen a rise in Islamic fundamentalism which, to my understanding, is largely a quest for power by certain groups and perhaps also a reaction to corrupt and inefficient government. Regardless of its rationale, the fact is that the Pakistani brand of fundamentalism is totally oppressive, if not brutal, towards women. 'Islamizing' the society means laws and restrictions which are horrific for women. The politicians, like the fundamentalists, are only interested in power, and one fears that they will become – indeed, are daily becoming – more and more powerful. One reason for this is the failure of the intelligentsia to take the issue of religion seriously. We have been so enthralled by 'progress', 'development', 'modernization' that the majority are ill-equipped to speak out against the fundamentalist vision of Islam. While we were preoccupied by Marx and development, power has been handed over by default to a bunch of thugs and terrorists. I believe that it is too late to stem the tide; the political and social agenda will be dictated by these people who use religion for power, and women here in Pakistan will be confronted with some profoundly difficult choices. 'Secularism' is not the answer – not just because the fundamentalist is not tolerant, but also because the secular/religious division is a false dichotomy.

The choice will be even more poignant and difficult because of parallel global developments. As I look around today, it is clear that my move towards a deeper understanding of religion and Islam was not an isolated phenomenon. My reading and my personal travels confirm that there is a sea change at work *vis-à-vis* religion. In Helsinki, New York, London and Lahore, people in general (and women in particular) are rediscovering religion. I say 'religion' deliberately because I do not think that there can be a substantive spiritual expression without the 'container' of a belief system – this is what a religion provides, or ought to, if it is to succeed psychologically. While the West searches for answers, more and more Muslim women are turning to Islam, not because of fundamentalism but because of what I can only call the present *Zeitgeist*; there seems to be a global phenomenon at work, not just in Islam. Thus, even as we recoil in horror at the 'Islamic' façade of fundamentalism, Muslim women are gaining a sense of the beauty of Islam. It is a strange dilemma: how to reclaim or sustain a religious life at a time when the religion itself is being made increasingly repellent?

My own situation is even more complicated, but may be paradigmatic of the future relationship between women and Islam – possibly other religions too. I belong (by birth) to a sect which has been legally classified as heretical in Pakistan; there is considerable persecution (even terrorism). To the extent that I believe that a re-visioning of Islam will come from Muslim women all over the world, the paradigm of heresy and its attendant persecution presents an exceedingly difficult 'choice' situation. This is my understanding of the future in Pakistan. Whether we want it or not (and I, for one, do not), religion will dominate our lives in highly oppressive ways. Yet, that same religion is vital to many, who see its oppression not as coming from God, but as *man*-made.

The choices facing Muslim women will be even more difficult, poignant (even dangerous), than those which others must tackle. Whereas most women for whom the spiritual is important may experience oppression or harassment from within their own dominant religious culture, Muslim women will find themselves besieged from within and without.

Islam and the West

There is no doubt today that many nominal Muslims (men and women) are turning to Islam as a response to what they see as a series of hostile currents originating in the West. The first of these is the Western need to fill the void left by communism as the 'Other', the enemy, the bad guys, with a new enemy – Islam. At the same time, recent geopolitical events (the Gulf War, Bosnia) have reinforced the realization that the West seems to be operating a value system essentially based on double standards – one for itself and another for the rest. While I am personally aware that not all Westerners or non-Muslims are anti-Islam, I am equally aware that in today's world the power of the media is enormous and that there is undoubtedly a very distinct image of Islam being circulated which is extremely negative. Over the last few years literally hundreds of editorials have appeared in newspapers across the Muslim world saying that fundamentalists are a threat to their own societies and not the West – but to no avail. One watches in amazement and disbelief at this continuing demonizing of Islam. There are certain values which one thought were universal, such as truth and justice, and which one believed were somehow more refined and evident in the West than in the 'underdeveloped' world. But the demonizing of Islam as Other/enemy is leading to a search for identity among those Muslims who have had exposure to Western systems of thought, now that the very people with whom they had so eagerly identified tell them that they are different – even evil – because of the religion into which they are born. This is indeed the stuff of self-fulfilling prophecies. The issue is compounded for women, who have benefited from the colonial fallout of education, for it has been thanks to Western education that women here have found a voice. For me, part of this educative process has been a reclaiming of religion and a reviewing of that Western system of knowledge which does not leave room for religion. Even as I see that Western women are themselves turning to (or at least searching for) spirituality, I find that I am being cast as the enemy simply because I am Muslim.

While this is highly simplified (inevitably, given the need for brevity), it raises this point: women in Pakistan will be in a state of siege from within and without, by a subject (Islam) which will increasingly dominate national and global politics. Whether we are looking at fundamentalism or the West, the main motivating force is not really religion, but the masculine addiction to power, in its various forms. In either case, women face stark choices.

4

WOMEN'S WORK AND FULFILMENT IN A CHANGING WORLD

Shanti George, India/The Netherlands

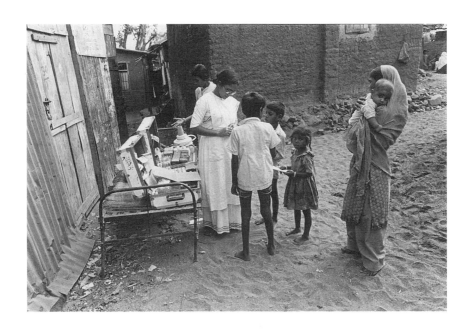

Mariamma, 1874–1970

Amoyi, 1903–1982

Leela, 1926–

Shanti, 1954–

Musings

April 1994: As I breakfast in our apartment in The Hague, I can hear my mother (who has come on a three-month visit from Kerala in southern India to help out) croon in the Malayalam language to my infant daughter, Anisa: 'Wake up, little one, roll up your sleeves and get on with a woman's morning work. Sweep the dead leaves from the garden, draw water from the well, steam rice flour and coconut for breakfast . . . '

In the bleak light of an early spring morning in The Netherlands, her words carry me far away to sunny, green Kerala. The world that my mother evokes is that of her childhood more than sixty years ago. In her present home in Kerala, water comes from a tap, and riceflour and coconut are steamed over a gas fire and not on the traditional hearth. These amenities have now been installed in the turn-of-the-century house where she grew up. My hope that Anisa will get to know this and other family houses well, symbolizes my desire that she should be well connected to her Kerala roots even if she grows up in Europe.

Little Anisa's breakfast will in due course probably come out of a cereal packet, but I hope that in the mornings of her life she will roll up her sleeves and get on with a woman's work. But how does one define a woman's work? How does it differ from 'work' as such? My mother's own morning responsibilities include her profession as a medical doctor, as mine do my work as a social anthropologist. What will Anisa's work be, when she becomes an adult in the early decades of the twenty-first century? Will she also participate in the traditional 'women's work' of reproduction and nurturance, as my mother and I have done, in addition to our professions? This 'women's work' links me to my grandmother and great-grandmother as well, both of whom I have had the privilege of knowing. What should I tell Anisa about these four female predecessors?

MARIAMMA, 1874–1970

A manager

Mariamma's life spanned a period of tremendous social change. She differed from her foremothers. She was literate in Malayalam, and to the end of her life enjoyed reading the daily newspaper. Mariamma's family was engaged in trade, in which she acquired some experience; she married at 18 which, at that time, would have been considered relatively late. Her husband was not only a landowner: he had a university degree and worked for the State educational service for a long time as an inspector of schools. His interests, in fact, lay in things other than land, and the management of their property was to a large extent in Mariamma's capable hands. After some time, he even transferred some property to Mariamma to enable her to handle it more effectively. In earlier generations, women whose husbands were primarily landowners would not have had such managerial opportunities. When her husband died relatively early, at 60, Mariamma was well able to handle the day-to-day management of the property, aided by her children.

Mariamma was the last woman in her line with little formal education (although the education she did have set her apart from previous generations of women). All her daughters studied to postgraduate level, and four out of five trained as medical doctors. The youngest of these gained a doctorate from Oxford and became internationally known for her work in anatomy.

Mariamma's marriage was – of course – arranged by her family. Of her five daughters and four sons, however, only one (my grandfather) had a conventional marriage arranged for him. The others married people whom they encountered in the course of their lives and work (one son married a Scot, another a German). One daughter, a busy surgeon, married only at the age of 40 and was childless. The youngest two daughters made what was seen at the time as a choice between their profession and domesticity, and remained unmarried. The youngest daughter (the anatomist of international repute) provided something of an alternative role model to later generations of women in the family, as she exemplified a life path other than domesticity. These later generations continued the tradition of female professionalism, many as doctors and scientists.

Much of the credit for Mariamma's daughters' achievements lay with their father, an educationalist closely associated with pioneering work in female education in Kerala. If Mariamma's role was a supporting one, it was none the less critical. On the one hand, she did not undermine her daughters' development by overtly or covertly expressing anxieties that this development differed so much from that of their female contemporaries (anxieties that would have been quite natural in a mother of the time). On the other hand, she was responsible for the

57

smooth articulation of the material base that supported the entire family. A grandson-in-law of hers describes Mariamma and her husband as a very effective team: he was a visionary and she was an astute manager who helped implement the visions.

Did Mariamma feel inadequate and unfulfilled, surveying the achievements of her daughters and the greater freedom that they enjoyed in their marriage choices? I suspect that she considered their achievements as her own. My mother (Mariamma's granddaughter, who knew her well) says: 'For a woman of her place and time, I think she found fulfilment.' Certainly, comparing herself with her own mother, Mariamma could feel that she had moved beyond older conceptions of female activity and women's work, thanks to her literacy, the scope she had for managing property as well as household, and the achievements of her children, particularly her daughters who were a generation or two ahead of their contemporaries. Mariamma's horizon of fulfilment had moved beyond that of her female predecessors, and this horizon would continue to shift for her daughters, granddaughters and great-granddaughters.

AMOYI, 1903–1982

A nurturer

Amoyi (my grandmother and Mariamma's daughter-in-law) belonged to the generation that followed Mariamma's; her life had more continuity with the previous generation than did the lives of Mariamma's own less conventional daughters. Like Mariamma's husband, Amoyi's father was not just a landowner: he had a university education and was a senior official in the State police force. Amoyi received more formal education than Mariamma had been able to a quarter-century earlier: she completed secondary school, although she married a year before she graduated.

Her husband (Mariamma's second son) worked for the colonial forestry service. When Amoyi was 30 years old and the mother of four children aged between 2 and 10, her husband was selected for special training at an institution in Edinburgh. A few months after he began studying there, however, he contracted pneumonia in the unfamiliar cold climate and died.

This was, of course, a turning point in Amoyi's life, and was partly responsible for the endearing combination of toughness and vulnerability that we, her grandchildren, came to know. Certainly she was the most vulnerable of the four women described here, but it could be argued that she showed the most strength in facing her life circumstances, if only because those circumstances were the most difficult. Two years after her husband died, she lost her eldest son to a typhoid epidemic at his boarding school.

Amoyi's anxieties about the future of her fatherless family were not borne out. The money that her husband had left proved sufficient (with careful management) to see one son into the army, and the other son and the only daughter through medical training. For almost a quarter of a century after her husband's death, Amoyi lived with Mariamma in Mariamma's rambling house, so as not to draw too heavily on the resources that her husband had left and to provide her children with the emotional security of the extended family.

In her fifties, after her children were settled, Amoyi used much of her financial reserves to have a pleasant house built in more than a hectare of garden and orchard. This was something that she had longed for over the previous decades: a home of her own, and a nucleus for her children and grandchildren that was separate from the main family house – and so it proved to be. In the very last years of her life, however, she may have felt that this home had come too late. When the family property was divided, one son (for understandable reasons) chose to continue living outside Kerala and the other son (also understandably) opted to inherit Mariamma's house where he had grown up. Amoyi's daughter already had a house of her own.

Amoyi was distressed at the uncertain future of her beloved house. This revealed the vulnerability persisting at the heart of the little fortress that she had managed to build for herself and her family. If she saw a woman's work as creating a home in which a family could thrive, and a woman's identity as expressed through such a home, she might have felt herself a failure now that the home she had created was bypassed. Had it not been for her husband's untimely death, the house would have been built at a time when it would have served as a long-term home to her children. If today she would feel distress at the fact that her house is rented out, she might find consolation in knowing that many of her grandchildren cherish happy memories of the family centre she created there during her lifetime.

Stereotypes of arranged marriages do not do justice to the deep affection that can unite wife and husband. The resources that her husband left behind provided materially for Amoyi and her children, but they could not fill the gap in her emotional life created by his death. She was a devoted mother and grandmother, and was in turn deeply loved by her children and grandchildren; but I think that there was a large and continuing emotional void that only her husband could have filled, and she outlived him by almost half a century. As a granddaughter who knew her in the last twenty-five years of her life, I was aware of how much her husband had meant to her and how deeply she missed him: she spoke of being reunited with him after death. There was thus an important part of her that remained unfulfilled: despite her achievements and fine qualities, therefore, Amoyi can be described as the least fulfilled of the four women described here.

I have often wondered if there was another area in which she was unfulfilled. With no more than school education herself, she took pride in her daughter's and granddaughters' educational achievements (three of her six granddaughters were awarded prestigious national science talent scholarships, an unusually high concentration in a single family). She was apparently content, like Mariamma, to achieve vicariously through later generations of women – yet she was a generation younger than Mariamma, and some of her contemporaries had benefited from university education (as, for example, did her own sisters-in-law, Mariamma's daughters). She never expressed any regret, but when I saw the enthusiasm with which she attended and took notes at Bible study classes, I wondered if she would have enjoyed some further education.

On balance, though, I think that if Amoyi's husband had lived she would have been a fulfilled woman, even without further education or professional experience (these were not within her horizon of expectations for herself). As with Mariamma, raising male and female children and grandchildren who became successful professionals was achievement enough for Amoyi. That she succeeded in this without a husband by her side might have appeared a significant achievement in terms of 'women's work', but for her it was a source of a lifetime of sadness and regret.

Whenever I am tempted to take short cuts in my research work – not to bother to check a reference or not to go back to field notes – I think of Amoyi's insistence that anything undertaken be carried out properly, and I remember the meticulousness with which she attended to the smallest domestic task. In this way, although Amoyi herself did no research, her influence on mine has been at least as strong as that of the more academically distinguished female relatives of her generation.

LEELA, 1926–

Family and profession I

Leela is Mariamma's granddaughter, Amoyi's daughter and my mother. Whereas Mariamma and Amoyi were women of their generations, Leela was ahead of hers. Unlike them she experienced 'compromises' in women's work. The source of certain compromises can be seen in her dual inheritance, one part of which included the example of her aunts, Mariamma's daughters. This example was clearly very strong: of Mariamma's eleven granddaughters, seven trained as medical doctors with various specializations, and Leela was one of the seven. The other part of her heritage was represented by Amoyi, who carried out so well the age-old female role of nurturance and reproduction.

When Leela thought of herself as a mother as well as a doctor, and looked around for models for this role, her attention was drawn to the one paternal aunt who was both a doctor and a mother. (Two other doctor aunts, as described earlier, were unmarried and another had married late and remained childless.) This aunt's elder daughter trained as a doctor, but the younger daughter declared (in implicit criticism of her mother) that she would be a full-time parent. Leela was deeply affected by this cousin's reaction. She had earlier thought of training further in surgery and specializing in plastic surgery. She now relinquished the idea as demanding too much commitment, at the expense of domestic life.

After her medical training, Leela married George, a young naval lieutenant. Their families knew each other, the two had met as children and as adolescents, and George was keen on marrying Leela. His family approached hers and Leela accepted. Soon after their wedding, George was sent to England by the Indian government and the couple lived there for a while. Leela felt distaste at what she read in the women's pages of the British press on 'getting your man' through artfulness and artifice: she thought that there was more dignity for women in the system of marriage that she was familiar with in Kerala.

Leela's later studies and employment were affected by the frequent transfers from place to place that George's work entailed. She trained in obstetrics and gynaecology as well as in paediatrics, rather than specializing in a single field. Between the ages of 26 and 40, she bore five children. The youngest child was born with Down's syndrome and required special attention and care.

Because of these circumstances (and 'compromises'), Leela could be described as the least professionally successful of the seven women doctor cousins. All the others specialized in a single field. One cousin who was close to Leela in age set up a flourishing maternity hospital in an Indian metropolis. Another married an army officer – but rather than follow him on his postings (as Leela followed George) she stayed in a large city with her children and built up a practice as an obstetrician. Leela, by contrast, opted for nine-to-five jobs, usually in government service, so that her early mornings, evenings and weekends could be spent with her family. Thus, not only was she unable to establish a private practice, but her work was in general medicine rather than in her specializations of gynaecology and paediatrics.

If Leela had to compromise professionally, further compromises were required in the domestic sphere. At a time when few naval officers' wives worked outside the home and most prided themselves instead on women's magazine-style housekeeping and entertaining, Leela's professional demands and five children prevented her from conforming to the model of hostess and socialite. She had, in any case, reservations about such a model.

In keeping her household going, Leela was often aided by her mother (Amoyi) who was not tied down by other commitments. Most notably, Amoyi lived for a while with Leela and George in England and took care of their (then) two children so that Leela could study and work, and she came to stay and help when each of Leela's children was born. The relationship naturally involved stresses. With her outside commitments, Leela could not meet Amoyi's meticulous domestic standards, and Amoyi was concerned about Leela's children growing up without full-time maternal attention. All the same, it was a deeply symbiotic relationship.

Despite these professional and domestic compromises, Leela considers herself a fulfilled woman, as she looks back on her life now that the demands of the home and the world have become less urgent. Indeed, she sees the compromises as trade-offs that have enabled her to combine professional and domestic fulfilment. She says that she has no regrets at having abjured her dream of becoming a plastic surgeon, at having accompanied her husband on each of his postings, and at not specializing in a single medical field.

She says that if she did not achieve certain things professionally, she achieved others. Although she did not work as an obstetrician for much of her life, she enjoyed the periods at the beginning and end of her career when she did so. She has also enjoyed the variety of work she has undertaken in the different places where she has lived. If she has earned far less than her cousins have from private practice, she says that she does not miss the money. Although she worked for long periods as a general practitioner in government service, she was able to use her training in gynaecology and paediatrics to extend her work to family welfare by providing services to employees' wives and children. She is pleased by reports that she is still held up as a role model to her successors in the job from which she retired. After leaving government service, she worked briefly as an obstetrician (to satisfy some lingering regrets) but then chose to work in occupational health with a large industrial concern. She is still active in occupational health, giving courses and taking up consultancies. She also contributes twice a week at a charitable clinic where she says that she is able to help patients who are near destitution. She is increasingly involved in programmes for children with special needs, after her experience with her youngest son. She was recently invited to work as a gynaecologist at a local hospital, but declined, saying that she has discovered that there is more to life than medicine.

Despite the stresses of a two-career marriage, Leela says that her combination of fulfilment both inside and outside the home depended heavily on George's support. At the beginning of their marriage he was not keen for his wife to take up employment, but he responded to her wishes and aspirations. At one point, when she returned briefly to medical college with a baby daughter and with Amoyi to help take care of the baby, he insisted that she take a house rather than stay with relatives, while he lived in naval bachelor accommodation. Leela and George's marriage has been a partnership of equals, with the storms that such a marriage can entail, and as such has provided a model for their children (including lessons on how to handle storms!). George's daughters find it hard to believe that he was once opposed to women working outside the home, so great has been his encouragement and his interest in their work.

Leela has attained professional achievements that Mariamma and Amoyi did not, and she is able to share their pride in having raised children who are achievers. In other words, she has sources of fulfilment that they lacked as well as the sources that they did have. (To what extent this double achievement of Leela's was underpinned by Amoyi's support is an interesting question.) Similarly, Leela experienced some strains and stresses which would have been familiar to them, as well as some to which they had not been exposed.

Leela is especially proud of the achievements of her youngest son, who despite Down's syndrome leads an active and productive life as an artisan and yoga instructor. She says that this has helped her to redefine achievement as it is conventionally perceived. Her most important remaining ambition is to help this son realize his full potential.

The gender relations in the family that Leela has raised are such that professional achievements and domestic capabilities are distributed relatively evenly between sons and daughters (one son cooks very well and has a magic shoulder on which restless children drift to sleep, and another is a skilled housekeeper – at the same time that both have done well in their professional fields). These gender relations have been strongly influenced by Leela's personality as well as George's: one son says, 'I'd like to marry a strong woman – I'm used to them.'

Perhaps what Leela says of her grandmother Mariamma could be modified for herself: 'For a woman who was somewhat ahead of her time and place, she achieved fulfilment.'

SHANTI, 1954–

Family and profession II

I am Leela's second daughter, Shanti. In early 1994, when I was almost 40 years old, my first child (Anisa) was born. Prior to my pregnancy, my interest in gender issues had been more theoretical and political than personal. The information that I was to give birth to a baby girl came halfway through my pregnancy, after the genetic testing that is normal at my age. The news focused my thinking on the implications of being born female into the everyday realities of the world in which I live: not just the global world, but my personal world. (This was also about the time that I became involved in the project on women's ways of perceiving and living their realities, and I appreciate the opportunities that this workshop has given me for reflection and discussion.)

The reason for this rather tardy personal confrontation with gender issues lies partly in the gender relations of the family in which I grew up (described above), with equal opportunities in education and the professions for women, and with men participating more and more over the years in domestic activity, especially as household help became scarce. There were thus no glaring gender inequalities to react against. This essay has so far dwelt on my mother's family, but my father's family too has many strong and capable women and a few professional achievers in the generation prior to mine. I grew up in what might be described as a benevolent patriarchy, with some matriarchal patterns that were seen as strengthening rather than undermining the solidarity of the patrilineal extended family. Interestingly, in my later career in India I encountered other benevolent patriarchs as research supervisors and directors of research centres, who provided me with professional nurturance. I wonder whether male egos in certain circles in India are so secure that they tolerate and even encourage professional achievement among daughters, sisters and wives.

Thus for me there were no serious gender battles to fight as an adolescent and young woman. One reason for this was clearly my class position. The women I have described here came from materially secure backgrounds, and their families were advancing further along the new frontier of the professions. This advance was surer if the women of the family also became professionals. Domestic lives could be underpinned not just by able family members like Amoyi, but by domestic servants. This is still the case to some extent, even if the 'servant problem' has become exacerbated in recent times.

Class lines were therefore in some ways more significant than gender lines, and once I left the little world of boarding school and moved to a college in a metropolis, I became increasingly conscious of skyscrapers rearing up over slums, of the poverty represented by pavement dwellers, of India's stark socio-economic disparities. In the vacations I participated in student camps for rural development that made me aware of rural–urban and intrarural disparities. Thus I entered the field of what is called 'development studies' because of expo-sure to socio-economic inequalities, and not because of experience of gender inequalities (as was the case with some women I know).

Within development studies, I chose to focus on rural development, and particularly on the production of a foodstuff – milk – that is vital to the well-being of very small children in some poor countries. I participated actively in the public debate over India's ambitious and controversial dairy development programme, and in 1985 published a book that criticized this programme. Earlier this year my second book appeared: it compares cooperative dairying in India and Zimbabwe, and examines how problematic cooperation between the different genders, generations and classes can be. The fieldwork on which that book is based was carried out over five years: it enabled me, among other things, to take a close look at women's lives in circumstances very different to my own.

Another set of issues that seemed more relevant to me than gender disparities during my adolescence were those connected to culture, since I grew up in a multicultural environment in which various strands of Indian culture were inter-meshed with various strands of Euro-American culture. In order to explore cultural issues more systematically, I took a Master's degree and a research degree in Social Anthropology. I have subsequently enjoyed teaching Social Anthropology at universities in India and Zimbabwe, since students there have experiences of (and interests in) multiculturalism that are similar to my own.

The previous paragraphs are meant to emphasize how tremendously fulfilling I find my work. I cannot imagine a life without it. This may appear in contradic-tion to some of the decisions and life choices that I am about to describe.

To return to an earlier line of discussion, if my great-grandmother and grandmother were women of their generation, and my mother was in some ways ahead of hers, what about me? To a great extent, generational differences have been ironed out. When I was at school and university, few of my peers had mothers who were professionals, but now many of these peers are themselves professional women and working mothers. Social patterns have changed, and some of the factors that influenced my family a couple of generations ago now operate more widely.

From some viewpoints, indeed, I am now a step behind my generation, having followed a life trajectory that must seem rather unmodern to some. At the age of 32, I was appointed as a Reader at India's apex institution for social anthropology and sociology, and would have had a good chance of going on to a professorship if I had kept up the momentum of research, publication and other professional activity. Instead, two years later, I resigned from this position (which was highly coveted by my peers and even by some of my seniors) in order to marry Anisa's father. He was then on secondment from an institution in The Netherlands to the University of Zimbabwe. I lived in Zimbabwe for a little more than a year, and have been in The Netherlands for the last five years. In these six years, my employment has been a series of varied assignments and no longer a tenured senior academic position. It seems unlikely that I will hold such a position again, especially now that I have new commitments of parenthood and have decided not to take up employment outside the home for at least the first eighteen months of Anisa's life.

After three generations of women in my family who have either kept abreast of their generation or have been ahead of it, do I represent a step backwards? I do not think so. I do not stand out among my peers as my mother and great-aunts did in their time, but by virtue of them I am a third-generation professional woman. Many of my peers, in comparison, are first-generation professional women and are often more 'driven' than I am. I do not feel under any obligation or pressure to plant flags on mountain peaks of female achievement for my family: those flags were firmly planted sixty years ago. In my own generation, my sister gained a doctorate in her mid-twenties: I could not better that and so had to find alternative means of achievement than taking a doctorate. If anything, one now creates ripples in the family through unconventional rather than conventional achievements.

Again, although I am not at present an achiever, I am an ex-achiever rather than a non-achiever. Having once risen quite high quite fast, I do not feel that I need to prove myself, nor do I yearn for high positions in institutional hierarchies. As with many other things, perhaps one needs to experience position and status before one can move beyond them. Similarly, I am often asked nowadays whether I miss 'working' (although I now work harder than I ever have in my life!), and especially whether I miss the 'jam' of academic life, the travel, the conferences . . . Having had my fill of these in the past (and hopeful of returning to them in the future), I am quite happy to do without them for the present.

I am also the third generation in a line of women trying to combine the work of nurturance within the home with professional engagement in the outside world. I trace my ancestry as a professional woman from the great-aunt who was both a medical practitioner and a parent, from whose life my mother tried to learn, and again from my mother who has attempted the same combination. I could have identified myself with other strands in my female ancestry: Mariamma and Amoyi represent the many women who found fulfilment through nurturance alone, and who created home environments that combined security with stimulus. On the other hand, my sister has chosen to align herself (as an unmarried successful woman scientist) with a third strand in our ancestry, exemplified by Mariamma's youngest daughter who gave priority to professional achievement. I am glad that my sister provides Anisa with an alternative model to domesticity. Why have my sister and I chosen the particular strands that we have? There is no space here even to try to analyse the combination of personality and life circumstances responsible for this.

My father said to me, when I was leaving my job and India, 'This break in your career need not necessarily be a break in your profession. Don't confuse the two. Life's ups and downs may affect your career, but your profession can remain steadfastly with you through a lifetime.' My decision was no doubt influenced by not just a mother but also a father who has always put relationships ahead of career considerations: as a result, neither parent has had a brilliant career, but both have had rewarding professional lives.

And in truth, while my career may have been in the doldrums during the last six years, my professional life has been expanded and enriched. Given the poor direct links between African and Asian countries, the travel opportunities in my earlier job were to northern America and western Europe; however, during my year in Zimbabwe I had the relatively rare opportunity to carry out a comparative analysis of an African country and India. An assignment a few years ago allowed me to conduct long, open-ended interviews with a variety of interesting people all over the world. More recently, I have spent a year discussing ecological problems with students from five continents. Sometimes when one is liberated from a career path, one's professional life can thrive.

It has been an interesting experience to be similarly liberated from institutional structures. For two of the last six years, I worked at home on a long book that was published in 1994. I was happy that my professional life had a momentum and a discipline of its own that was not dependent on institutional underpinnings. The invitations that I quite frequently receive to speak or lecture on various occasions lead me to feel that I am invited because of what I have to say and not because I occupy X position at Y institution. Indeed, over the five and a half years before Anisa's birth, I have always had some assignment or other, if not a single stretch of employment: these assignments mostly came to me and I have never had to solicit work. Thus, if I have moved away from institutional structures and conventional career paths, it is not for lack of a professional identity but because my professional identity is strong enough not to depend on institutions and career paths. During this period, though, some of my assignments have involved attachment to an institution. I enjoyed participating in the work life of a variety of such institutions, but was glad that these were sojourns and not long stays. Institutions are perhaps most congenial to work in when you can move in and out of them (rather than being penned within).

Most recently, after Anisa's birth, I am relieved that I can enjoy the early years of her life without pressures from an institution of employment. This chapter was written in brief fragments, in the interstices of giving a lively and active child the attention that she needs, and without recourse to day-care centres or babysitters. It is a simple, descriptive chapter, but I am none the less happy to be able to weave very different kinds of rewarding work into my life. I do not see the situation in either/or terms. Some years ago, when I was engrossed in writing a book, I heard about a friend's pregnancy and wondered if I would rather be having a baby. My immediate reaction was 'No.' Today, when I am taken up with caring for Anisa, if someone were to ask me, 'Wouldn't you rather be writing a book?', my answer would be another immediate 'No.' And I optimistically look forward to a future when I can be both Anisa's parent and a productive writer.

While I would not agree that my life choices represent a step backwards from some feminist agenda, I shall not try to present them as a step forwards. If anything, I have taken a step sideways, to explore new choices and combinations in addition to familiar ones, and to move away from notions of 'forwards' and 'backwards' on some sort of linear scale (where 'forwards' often connotes individual achievement through competitiveness). Neither do I see my life choices as 'feminist' in any narrow sense. I think that a man today can also choose to try to combine relationships with individual achievement, and nurturance of others along with the development of one's own potential.

I do not know what the future holds for me, but I hope that Anisa will be able to say of me, 'As a woman who lived in challenging times and places, she achieved fulfilment.'

REFLECTIONS

The four women's lives just described suggest that rather than 'finding' fulfilment within their environments, women 'achieve' it through active negotiations and confrontations, by articulating the opportunities and dodging or pushing back the constraints. They are protagonists who live by their own choices as well as by the choices of others, in their time and place. Whether they 'compromise' or 'trade off' is a matter of perception. At no point have I used the word 'compromise' when describing my own life.

What of Anisa, who is now a gloriously androgynous 14-month-old? What, if anything, can be projected about her time, her place, her interactions, her negotiations, her confrontations?

Some people say, 'Oh, you don't have to worry about Anisa, if she grows up here in the [presumably enlightened] West.' I smile, thinking of the pink envelopes that poured in at her birth, bearing cards that sported rosebuds, angels, dolls, kittens, lockets, ballet shoes, beribboned prams, lace handkerchiefs, frilly dresses – in short, sugar and spice and all things nice. It will be interesting to observe, over the coming years, the various elements of gender conditioning that operate in Anisa's western European environment.

I suspect that, as for me, gender in itself will not be an overriding preoccupation with Anisa. As an 'only' child, there will be no competition for family resources and no brother to set off overt or covert gender discrimination (as I have seen happen 'even' in western Europe). I trust that she will find her parents' marriage the (stormy!) partnership of equals that I perceive my parents' marriage to be, and will find no goads there to feminist indignation.

A difference from my formative years – if Anisa grows up in The Netherlands – may be that she will be less conscious of class disparities than I was as a girl in India, or as her father was as a boy in Britain, given The Netherlands' more even patterns of income distribution and its welfare state mechanisms.

Will racial identities make themselves felt in Anisa's life, in contrast to my own youth in a subcontinent where the tremendous physical diversity among its inhabitants is yet largely contained within a single racial framework? I wonder. I feel comparatively at home in The Netherlands because of the cosmopolitan ease with which racial differences seem to be treated, at least among most people with whom I come into contact. Is this – as some suggest – the smooth opaque surface presented to a relatively late entrant who is seen as a transient? Will Anisa's experience be different, as she grows up within the complex social reality that pulses under the smooth surface? If so, she will be perceived not only as a member of a racial minority but as a female member, since race and gender are usually closely interwoven. On the other hand, will her class position and/or membership of a marginal circle of expatriates shield her from racism?

I am fairly certain, though, that she will be made strongly conscious of cultural diversity. There will be some important differences from the culture-consciousness of my own youth, mentioned above. In my childhood, cultural differences were striking, piquant, even mutually contradictory, but they were challenging rather than threatening. Strands of different cultures were fairly closely interwoven in daily life. Like most Westernized Indians, we were north Indian/south Indian in some things and Western/Anglicized/Americanized in others, switching tracks or mixing them where this was felt necessary, and not consciously thinking about these things most of the time but just getting on with our lives – just as we would switch between an Indian language or languages and English, depending on the context. Sometimes we did think about and discuss why we did particular things in a certain way. Readers from other ex-colonies will probably recognize these patterns.

While I will try to reproduce this routine multiculturalism in our domestic world, the world that Anisa will encounter outside is less integrative of cultural diversity. Contemporary Europeans do switch cultural tracks in their daily lives, between various continental European tracks and Anglo-American tracks (and the Dutch seem particularly adept at this). Their set of tracks is, however, far more restricted in variety than are those of Westernized Asians, Africans and Latin Americans. Thus, part of Anisa's cultural world (or multicultural world) will be alien to those she lives among: some of those she interacts with will see this not just as alien but as threatening, and many will tend to cast the alien in frameworks of inferiority/superiority. She will be called upon to describe, interpret, explain, justify, mediate: no doubt, at times this will be interesting and challenging, but there will also be occasions when this will be tiresome, even infuriating.

Much of this, inevitably, will be bound up with gender: 'You're part Indian? You're lucky to be living here and not there. Women don't have much of a life there, do they?', or even 'Girl children are killed in India, aren't they?'. At the very least, Anisa will have to tolerate such exchanges, and perhaps endeavour to delineate something of the complex realities of India. (Given the family background described in this chapter, it will be difficult for her to go along with the stereotypes.) In order to do this, over the years she will have to think through and live through various aspects of her cultural identity, her gender identity and the combination of these. It is possible that she will simultaneously have to think through her racial identity and her class identity, and how these interpenetrate her gender identity. These identities, and the configuration they form, will inevitably be fluid: they will change, mutate and metamorphose over the years as relationships evolve and new relationships emerge.

And yes, somewhere entangled in all this, and perhaps at the most conscious level, she will have to think about her work identity, not just about who she is (a woman, half-Indian and half-British, probably resident in The Netherlands), but about what she is to 'do for a living'. Will she be a social scientist, like her parents? Or work in the pure and applied sciences, like most of her mother's family? Or will she break with tradition, and 'do her own thing', whatever that may be?

These are questions of career and profession, but the lives of her female predecessors suggest that these questions will be intertwined with larger, more shadowy, less overtly conscious life questions – the questions raised by women's life circumstances in particular times and places, and the questions that women themselves ask in the face of these circumstances. What does it mean to be a woman in my context? What space can I find for myself within these definitions of what a woman is and what a woman's work is? In what ways do the definitions constrain me? As these definitions can change and are changing, how can I move towards a redefinition that is closer to my own perceptions and aspirations?

NOTE

The four lives that I have described are part of the wider story of the Syrian Christians of Kerala. Kerala is often mentioned in the 'development literature' as an example of development defined in terms other than economic growth, including the status of women:

> During this century, Kerala's people have had quite remarkable success in improving their health, raising their levels of literacy and education, and bringing down the birth rate. And all this at a level of per capita income that has been lower than that for the country as a whole. Recognition of these achievements by policy makers and researchers has led to a view of Kerala as representing one type of 'model' for social development . . . Kerala's early inroads in education, particularly female education, gave it a head start which the rest of India has not caught up with today. Clearly, Kerala's performance owes much to the fact that the female half of its population has not been left out in the cold.

(Sen, 1992: 254, 268)

Others (such as Robin Jeffrey, 1993) suggest that Kerala is not a 'model' but does offer some important lessons about women and development. K. Saradamoni warns sharply against complacent and uncritical presentations of a 'Kerala model'.

Kerala has larger non-Hindu communities than elsewhere in India. Situated on the Arabian Sea, it has a long history of peaceful conversion to Christianity and Islam through trade rather than conquest, and an amicable record of communal relations compared to other parts of the country. The Syrian Christians are an influential minority community: 'The Syrian Christian Church is one of the oldest in the world and dates back to the first century AD . . . The growing prosperity of the Syrian Christians in the nineteenth century generated a demand for education' (Sen, 1992: 264). The shared religion encouraged many Syrian Christians to make use of the educational institutions set up by British missionaries. Education in turn enabled the community to extend its sphere of influence from trade and land to the professions. The stories I have told, embedded as they are within the historical and cultural c ntext of the Syrian Christian community, and part of Kerala's history of development, illustrate some implications of this for the women of the community, over the generations.

REFERENCES

Jeffrey, R. (1993) *Politics, Women and Wellbeing*, New Delhi: Oxford University Press.

Saradamoni, K. (1994) 'Women, Kerala and Some Development Issues', *Economic and Political Weekly*, 26 February, pp. 501–509.

Sen, G. (1992) 'Social Needs and Public Accountability: The Case of Kerala', in M. Wuyts, M. Mackintosh and T. Hewitt (eds) *Development Policy and Public Action*, Oxford: Oxford University Press/Open University, pp. 253–277.

5

THE ACT OF BREATHING

Esperanza Abellana, Philippines

Ilang de la Cruz, 1914–

Feny Paradela, 1926–

Fernandita Abadia, 1950–

Iris Paradela, 1973–

ILANG DE LA CRUZ, 1914–

A lifetime of hard work and faith in God

At 80 years of age, Ilang de la Cruz is actively involved in all the chores in the house which she shares with her youngest son, a friendly man of about 40, who has Down's syndrome. To earn extra money, she has rented the upper part of her house to some young women who have come to the city to work as salesgirls in the department stores. Her house is located on a privately owned lot; although she and other families have lived or 'squatted' on this lot for a number of years, they constantly fear the threat of demolition. In spite of her age, Ilang is rarely at rest. From morning until night, she is busy with a wide variety of activities: marketing and selling vegetables, taking care of her pigs, going to church, doing apostolic work as an active member of a church organization, attending meetings or doing some kind of community work as a leader in the Basic Christian Community of the parish. She is also a member of the urban poor community theatre group that performs in the church, in communities and on the streets during rallies and other occasions.

Ilang has demonstrated strength in her convictions as well as physical strength. She was born and raised in the mountains of Takay, a barrio of the town of Sogod on the island of Cebu in the Philippines. Her parents, who were farmers, used to leave their nine children at home while they worked in the fields. Being the eldest, she was given the responsibility of looking after her younger brothers and sisters. From her early schooldays, she had to prepare meals as soon as she came home from school. She recalls her childhood as a time of all work and no play: her chores included fetching water from the well, cooking, feeding the chickens, putting the carabao into the shade and digging sweet potatoes. For Ilang and her brothers and sisters, failure to do their chores meant a severe beating. She was not allowed to say anything when punished, for this was considered fighting back. She feared both parents, but especially her father, for his beatings were worse. Whenever her father beat her, she would kneel down in front of him and beg to be spared from the whip.

Ilang remembers some of the 'house rules' from her childhood: never to be around when adults were talking, and not to join *barkada* or groups of close friends. Her mother taught her to pray in Spanish and her grandmother forced her to join in daily prayers at 4 a.m., 6 p.m. and after dinner. Since she was too young to understand what praying was all about, this became just another chore for her. In spite of what she went through as a child, she claims to be grateful, because she learned to work hard and to take responsibility, and she developed a strong faith in God. Raised in the tradition of regarding obedience to one's parents as a primary obligation of children, Ilang learned to accept her situation without developing even the slightest negative feeling towards her parents.

Her parents made her continue her education up to first year of high school. Since there was only first grade (i.e. the first year of schooling) in the barrio, she had to go to school in another town. She lived with her grandmother; at weekends, she would go home to the mountains to prepare her food for the whole week. After finishing her first year at high school in the city, she began work as a dressmaker. Looking back, Ilang believes that her parents instilled in her the value of education, which later made her strive for her own children's education.

At 26, Ilang married a man who worked as a driver. It had taken ten years before she would agree to marry him because she wanted to be sure that he had a job. Some time after the marriage, she discovered that her husband had in fact been carrying on a relationship with another woman while engaged to her: although he behaved properly as a husband, she still felt cheated.

She bore nine children and became a full-time housewife. Since she could not afford to hire any domestic help, due to her husband's meagre income, she trained her children to do housework. She was soon exacting from them the same strict discipline which her parents had expected from her.

In 1974, her husband died, leaving her without any means of support. Her main worry was how her children would finish their studies and get their degrees. She felt that God was sending her a message and her task was to accept the challenge. At this point, she realized that, having lost the feeling of security which her husband provided, she was forced to draw from her own strength – *gikan sa kinahiladman sa kaugalingon*, literally 'from the deepest part of oneself' – to be creative, resourceful and courageous. From then on, she worked even harder and struggled to scrape together enough money for all but two of her children to get college degrees. Their success gave her enormous pleasure and a sense of fulfilment, especially when one of her sons became a Catholic priest with the help of some sponsors. Now her security is that her children come to her rescue in times of crisis.

Ilang has always been considered a leader in the community, particularly in church and parish activities. Her religious involvements have always taken priority in her life because through these she can manifest and express her commitment to serving Mary, the mother of God. The feeling of contentment which this gives to her is a carry-over from the religious training which she received as a child. As a member of the community theatre, she is also aware of social realities; mainly through drama, she has been actively involved in protests against the Marcos dictatorship and militarization (even though one of her sons is a military man), against social injustices, and human rights abuses.

FENY PARADELA, 1926–

A clash of consciousness

Feny is a soft-spoken woman who looks older than her 68 years. She works as a master-cutter in the sewing shop of a college run by Catholic nuns. She lives with her husband and two children in the squatter area. Feny, who was born into a family of eleven children, grew up in the mountains of the town of Naga, Cebu. Both her parents were farmers, although her mother mostly stayed at home.

In 1949 a typhoon destroyed their crops, including all the coconuts which were their main source of income. From then on, Feny's life changed. She was in sixth grade (i.e. sixth year of schooling) when her parents asked all the girls to leave school; the boys were allowed to continue, although in the end only one persevered with his education. By the age of 13, Feny was working as a salesgirl in a small shop in the centre of the town and contributing financially to the education of her elder brother. They had agreed that once he finished high school and found a job, he would in turn support Feny: her brother fulfilled his promise and later paid for her to take a course in dressmaking.

At 17, Feny was secretly married. It was done in secret because her parents were opposed to her marrying this man, the husband of her deceased aunt. Since she was not yet old enough, she needed her parents' formal consent to make her marriage legal. Some years later, her father and her husband had a major disagreement and her husband made her choose which side she would take: she chose her father and left her husband. She lived with her parents and worked as a private dressmaker to a rich family, frequently leaving her son, Victor, in the care of her parents. When her son was 12 years old, Feny married Isagani Paradela, a man who worked at the registrar's office of Don Bosco School where her son was enrolled. Again her parents did not approve of the marriage, because the man came from a broken family; but again, Feny made her own decision.

Three years after their wedding, Feny found that her husband was seeing other women. When she confronted him about it, his excuse was that this was his only form of relaxation, since he neither smoked nor went drinking. Even after their first child was born, her husband continued to have affairs. Feny lost her trust in her husband and fights between them became a common occurrence: she used words and he used his fists. On several occasions, Feny told him to leave the house and not to return, but he took no notice. He continued to come home and to make her life miserable. Then he was forced to resign from his job because of a scandal created by two women who were fighting over him in his office. Even when given the chance of another job he refused, saying that it would give him the chance to become involved with other women again, so he would rather stay at home.

Fortunately, Feny found a job in the sewing department of an all-girls college, which meant that she was able to support the family. While she was at work all day, her husband did the household chores and took care of the baby. Even with this arrangement, life was hard for Feny; when she finished work in the afternoons, she would still have to go to market and then prepare dinner when she got home. Things got worse when Feny became pregnant with their second child. Her husband was angry about the pregnancy, claiming that she should not have become pregnant while he had no job. He gave her money to pay for an abortion, but Feny fought to keep the baby.

Being the sole breadwinner with two children and a husband who is a burden to her, Feny feels that she has had a very hard life. She feels that she has been very unlucky (*walay suerte*) in her married life. Behind this feeling, there seems to be a certain degree of fatalism, that things are beyond her control and that all she can do is hope for good luck. She describes her relationship with her husband as a 'clash of consciousness' (in Cebuano, *sumpaki sa kabubut-on*). However, she still sees her two good children as a sign of God's mercy and kindness to her.

At this point in her life, Feny feels much freer. Last March her eldest child finished college with a special award for campus leadership. Her husband refused to attend the graduation; her second child is presently a full scholar at the University of the Philippines. Her children are her main source of inspiration. She says that she and her husband have agreed to separate after their son finishes his studies.

FERNANDITA ABADIA, 1950–

The value of love and respect

Fer, as she is popularly known, is 44 years old and a teacher. She is the eldest of four brothers and one sister. Her father was 60 years old when he married her mother, who was then 27. As a child, Fer used to hear her mother say, 'Your father has almost become a hunchback from working', which made her keenly aware of her father's sacrifices for them. She realizes that her mother taught her to love and value her father. Fer remembers her father as one who loved his children very dearly. She recalls that when they were still small, he would choose to eat only vegetables at mealtimes so that the children could have the meat or fish.

As a young girl, Fer stood out among her classmates in school because she was tall and pretty. Although not conscious of her good looks, she was aware of being tall, which made her shy. At the age of 11, she and her classmates were taught to dance *Pandanggo Sa Ilaw*, the dance of the light. This is a Filipino dance in which the dancers perform in a darkened place or room, with lighted lamps on each hand and on their heads. Fer danced so well during practices that the dance teacher told everyone to look at Fer as the model dancer. This extra attention helped her to develop more self-confidence. But during the final performance Fer's lamp fell and broke while dancing; it only happened to her, everybody else danced perfectly. This was a huge blow: she was ashamed of herself for not meeting her teacher's expectations and she was disappointed in herself. After that, she withdrew from her friends and lost confidence in herself once again.

One person who influenced her greatly was her father's adopted son, Juanito. Although he was adopted before her father married her mother, Juanito was like an older brother to Fer. As they grew up together, Fer noticed that her mother did not treat Juanito fairly, sometimes even wanting him to leave the house. Fer felt this to be unjust, so she asked why he was being treated that way. It was only then that she learned he was not her real brother. What impressed Fer most was the way Juanito took everything without ill-feeling. He continued to be obedient and caring to everyone; he did not fight back or show signs of wanting revenge on her mother. Because of her feelings for him and her respect for his attitude, Fer continued to love Juanito even after she knew that he was adopted.

Then something happened that changed her mother's attitude towards Juanito. Her mother had an ectopic pregnancy, and came close to death through blood loss. The type of blood she needed was a rare type, and by chance Juanito had the same blood type. As well as providing the blood, it was Juanito who looked for transport for her mother to be taken to the hospital. Fer's mother survived and, after that incident, became very good towards Juanito. Fer is full of awe for Juanito as she recounts the story. She is proud of him and looks up to him with love and respect.

Fer is happy in her married life. She feels that her husband is a real partner who complements her weaknesses. Her husband does most of the cooking, because he is a better cook than Fer, but he has never raised this as an issue against her. She is deeply grateful to him and thinks only of what she can do for him in return. Although there was a period of ten years when her husband had no job, which meant that they were hard up financially, Fer did not make this a problem. She felt quite at peace – even then – because her husband had shown her how to give and take. None the less, Fer is not comfortable about telling her husband that she gives financial support for the schooling of Juanito's adopted child, as well as providing financial help to her other brothers when they are in need. She feels that her husband might be hurt if he knew, because she feels that her primary responsibility is to her own family.

In their relationship, Fer and her husband rely a lot on God's presence. They both believe that in any decision-making God is also with them. As a way of acknowledging God's presence, which they value so deeply, they attend Mass every day.

Fer realizes the great influence her mother had on her in terms of the love and respect she felt for her father, and she has unconsciously trained her children in the same way. When asked whom they love more, her children's answer is their father. This makes Fer feel happy and satisfied. For her, it is important that she can see and hear the way her children value their father. Being their mother, she feels completely secure because 'they have come from my own flesh'. She wants to share this feeling of security with her husband through the children.

As a mother, Fer also makes sure that her children learn to experience God. She goes to the prayer room of Santo Rosario Church every Saturday afternoon and she makes a point of bringing one of the children each time. She tells her children that 'This is the place where mommy comes to tell her problems.' On one occasion, her 9-year-old son broke a light-bulb in school. His parents were called to school but the son asked Fer not to tell his father. Fer could not agree to this, but said instead, 'Ask Jesus and Mary first and we will see what to do.' Much to Fer's surprise, the boy responded immediately that he had already done that: he had gone to the prayer room of Santo Rosario Church by himself. Fer felt a deep happiness that her son was beginning to experience the reality of God. She wants her children to grow up believing that even without their parents they will always have a companion in their lives.

Fer's world is very much confined to family and school concerns. She enjoys socializing with her fellow teachers in school and with family friends. She has her own opinions and enjoys taking part in discussions on social issues and realities (including women's issues), but she has not joined any formal organizations.

IRIS PARADELA, 1973–

'Women can lead and make a difference'

Iris is a frail-looking girl of 21, who graduated from an all-girls college just a year ago. Being the daughter of Feny Paradela, who works in the sewing centre of the school, she was able to study without paying the tuition fees.

She has been raised in an unconventional home in the sense that her mother goes to work while her father takes care of the home, instead of the usual 'father/breadwinner', 'mother/homemaker' arrangement. Because of this, she has developed a different attitude towards the role of men and women at home and in society. Iris sees her mother as the leader at home, the one who usually makes the decisions. In relating with her friends she finds herself asserting her opinions, and claims to have learned this from her mother.

Iris has one brother who is now a college student at the University of the Philippines. When they were still children, her brother tended to follow her around, play with her and share her toys. He grew up to be gay, which Iris attributes to the lack of a traditional male model in the home and to her brother's acquaintance with some gay students at school. At first, it bothered her because gays have a difficult time in society, but now she has accepted her brother's orientation because she sees that he is happy.

She believes men to be egotistic. In her relations with men, she has the urge (and the tendency) to point out their weaknesses just to prove that they are not superior to her. She finds this satisfying, feeling that her achievements must be greater than those of her boyfriends. At the same time, she is apprehensive about reaching out to men because her actions might be misinterpreted. She acknowledges her negative attitude towards males and realizes that her personal experiences are behind it. She makes an effort to check herself by consciously putting aside her prejudices and preconceived negative notions while interacting with men.

In her childhood, Iris experienced a double standard which made her angry: her cousin (a boy) was allowed to play outdoors with their neighbours, while she was allowed neither to play with them nor to leave the house. Looking back, Iris still thinks that this was unfair. She also recalls her mother advising her to be cautious about the people she associated with, and to avoid making friends with rich classmates. Although she heeded her mother's warning during her early years in the elementary school, she later ceased to be conscious of it: being poor did not prevent her from becoming close friends with rich classmates because she felt accepted by them and learned from them that the rich had problems too.

In general, Iris is fairly self-confident, especially when she knows that there are people who can help her or whom she can turn to. She is also an optimistic person, and says that in this she must take after her father, who always believes that things will turn out right. She places importance on having strong faith in God, feeling that it is this religious faith that will give direction to her life. For Iris, spirituality is not confined to church, because she claims that religious leaders are often unable to inspire others or to serve as role models.

During her college days, Iris was an active student leader, receiving a certificate of merit for campus leadership during graduation. Her experiences as a student leader made her realize that she could take responsibility and that she could influence people. She is firm in her belief that women can lead and can make a difference. On the other hand, she is not inclined to get into discussions on political and social questions. For her, the events of 1986 at EDSA (the four days of 'people power' which led to the downfall of Marcos and the end of dictatorship) were a significant affirmation to the Filipino people of what they could achieve for their country. At the same time, she does not believe only in activism and criticizing the government: she feels that it is now time to think and to work out creative new alternatives.

Iris's immediate goal is to find a job so that she can earn money; she wants to help pay for the education of her cousin, a burden presently shouldered by her mother. Being a new graduate, she is eager to make herself useful, but she wants to do something 'different', something that has an impact on society, although she cannot yet define what exactly that will be. As a graduate in Psychology, she sees herself as one who helps, leads, inspires and encourages other people to achieve their full potential so that they might one day influence society, as policy-makers or politicians, or in other occupations in line with their field of specialization.

REFLECTIONS

The women whose life stories are reported here were chosen for a number of reasons. First, I wanted this study to reflect the self-understanding of a peasant woman, an urban poor woman and a woman professional coming from the sector of the rural and urban poor. Second, I felt that the process which these individuals would undergo as they reflected on their lives as women would be a liberating experience. I therefore chose women who have a circle of influence outside the family and who would be most likely to share their experience with others. Third, this reflection on four lives was offered in the hope of contributing to a better understanding of ordinary Filipina women.

My acquaintance with most of the women can be traced back to a period of seven years when I was a community organizer of Basic Christian Communities in the area where most of the women come from. Only Fernandita does not live in the urban poor community. Feny Paradela is the mother of Iris, but there is no relationship among the other women; neither am I related by family ties to any of them.

Reflecting on these four life stories, I am struck by the image of an ocean wave. Coming from a small island, Cebu, I am constantly exposed to the ocean; in a way, my life is entirely connected with it, since I can neither leave nor return to the island without taking account of the ocean.

The identity of a woman cannot be separated from her experiences of fulfil-ment, struggle and crisis. This is the context of her 'wholeness' as she responds to personal challenges in relation to family and community. Like the wave in the ocean that appears visible and distinct as a wave-form for just a moment, the woman's individual identity is in constant motion – rising up briefly and then being submerged again with the shared identity of the people with whom she interacts. In another moment the ocean wave will blend itself in ever-widening ripples with the foam and the waters. The situation is never static. Sometimes, the waves are bigger and more defined; sometimes they are small and indistinct.

There are many common threads interwoven throughout the stories: respon-sibility and sacrifice, creativity and tenacity in achieving one's goal, a strong spiritual or religious sense, the willingness to serve the interests of family members. All the women share the same context of poverty and family relation-ships; they hardly spoke of relationships outside the family. Perhaps they felt that the experiences to be shared should be limited to the family, or perhaps, for these women, their personal identity and experiences of 'wholeness' and fulfilment as women are those that directly relate to the family.

The ordinary Filipina is generally trained to feel responsible for her family. As a young girl, she grows up with the constant reminder of a particular behaviour and a set of duties and responsibilities which she is expected to fulfil. During childhood, parents are often more tolerant with boys than with girls: boys have more freedom, more time to play, more opportunities to get what they want or need. Thus, even as a little girl, the socially constructed self-image of the Filipina – which may overshadow her own idea of self – is already being formed, consciously or unconsciously.

The sense of being responsible for one's immediate family remains even after a woman establishes a family of her own. If she finds a job, she is expected to think of her parents, her brothers and sisters, and any other relatives who may need help. Indeed even if she doesn't have a job she must still remember them. In this regard, there is little or no differentiation between rich and poor women: although the burden naturally falls more heavily on the poor, both go through the same experience and share similar attitudes about themselves and their re-lationships with their family.

To a great extent, the economic and political system of the Philippines promotes the worsening conditions and the difficulties faced by poor women. As the country develops an increasingly money-oriented economy, lifestyles and almost all other needs are money-based. This can lead to heavy demands being made on the women, for more often than not it is they who look for ways to make ends meet, even if it means sacrificing themselves. If the basic needs of the family are not being met, the woman strives for access to additional resources – hence the growth of the 'informal economy' that sustains the lives of a majority in the urban areas, especially women and children. Many women leave their families to join the overseas contract workers (OCWs), a government scheme expected to bring in foreign exchange. A 1995 report from the Philippine Overseas Employment Administration and Department of Foreign Affairs shows that 60 per cent of the 700,000 Filipinos involved in this scheme are women. The phenomenon of mail-order brides and the number of women leaving the country to work as entertainers are, either directly or indirectly, manifestations of the same reality. These women will do anything possible, even at great risk to themselves, to meet the increasing financial needs of the family.

The pressure of survival and the economic environment in the country push women into different arenas of tension. When confronted by a crisis, it is the women who sustain the family while finding ways to resolve the crisis; the men usually become paralysed and unable to act. At the same time, the woman also has to find ways to ensure her own survival, so that she can continue to fulfil her role as mother and wife. In the Philippines, where very few have access to any form of social security, the family has become the only system which can provide emotional, psychological, moral and economic support. Much of this support is provided by women.

Among the urban poor organizations, it is not unusual for women to become active members or leaders. In spite of the fact that these women have been busy all day, not only with household chores but also with the work they engage in for income, they still attend community meetings or seminars in the evenings. To them, it is a matter of balancing priorities – attending both to domestic obligations and to the needs of the community. The men usually excuse themselves, because they have to relax and drink with their friends after a day's work. In the Church, in government and in the social movements, women have taken up significant roles: today, there are at least fifty women's organizations in the Philippines. The ability to forget one's individual interests in favour of a common commitment is an experience shared by many women in the country. Herein lies the woman's strength as support and as leader in the family or community.

Filipinos look towards the family for motivation, for a sense of meaning and loyalty. Filipino society is said to be strongly family-centred or family-oriented, which means that the interests of the family usually come before any other consideration: whatever is best for the family is identified as one's own best interest. Many of the women who take on so much responsibility for their families fuse their own identity with that of the family. It is not surprising, therefore, that few Filipina women consciously consider themselves as separate from their husbands and children. Identity is not individual, but shared: a personal decision is usually taken within the context of this shared identity, either with parents, brothers and sisters, children or other relatives. The attitude that individual identity is only good to the extent that it contributes to the total well-being of the family is especially strong among women. They form an identity which is based entirely on their relations with others, on their ability to respond to the needs of a situation or a person.

The woman's feeling of 'wholeness' thus lies in the extent to which she can keep a balance between her individual and shared identities, maintaining the process of submerging and emerging, like the ocean wave. Even if she spends most of her life 'submerged' in this way, the important thing for a Filipina is that she experiences a positive feeling in her *loob*, her inner self. On a recent train journey, a Filipina bound for work in Saudi Arabia broke down in tears when I asked her where she was going. After she had recovered, I asked her why she cried; she told me of the two little boys and husband whom she had left behind. But she quickly said that she must have *lakas ng loob*, a strong inner self, for the sake of her family.

Loob is a key word in the language: a person's disposition can be described according to her/his *loob*. *Kagandahang loob* refers to a good disposition, and *masamang loob* to ill-feeling. To a Filipina, I believe that the gauge of personal fulfilment is the condition of her *loob*. There is a slight difference here with the Cebuano language: in Cebuano, it is not *loob*, but *buot*; and *buot* refers to two things – consciousness and will. Consciousness is something inside the person, like *loob*, but there is also the sense that this can, to some extent, be influenced by the will of the person.

One can illustrate this concept from the life stories: Fer feels happier when her children express their preference for their father. She does not seek assurances for herself, because she bore their children, but she feels that her husband may need this assurance. According to Fer's perception, her self-identity is established through motherhood and is thus very secure; her primary concern is to pass on the feeling of security to her husband. This example also seems to reveal the significance of motherhood to Fer's sense of individual fulfilment and self-confidence. She consciously creates the kind of situation that she wants to promote in the family. When she succeeds, she is bringing about *maayong buot* (the disposition of a good person) or *magandang kalooban* (beautiful inner self) to her husband and to herself.

Lack of fulfilment or wholeness occurs when a woman is in a situation of *sumpaki sa kabubut-on*. This means that her decisions, feelings, convictions or principles are not reconciled with those of the person with whom she is closely interacting in a particular situation. It can also occur when the woman cannot act on her own choices, or follow her *buot*, because of certain circumstances. In the life story of Feny, for example, she has worked hard and made sacrifices all her life for her family, with two contrasting results: fulfilment from the children and frustration from the husband. She tries to keep herself in balance while she pursues her dream for her children.

Women's spirituality is another thread that runs through the life stories. There seems to be a connection between the social role of women and the fact that, in the Philippines, more women than men attend religious services and frequent religious places. Perhaps their spirituality is based on their need to implore for added strength and power: one of the most heavily attended church services is the novena to the Mother of Perpetual Help held every Wednesday. Other religious practices that attract a lot of people are those that involve penitence, such as the 'dawn rosary' and 'miracles'. In all these events, the number of women participating is striking. These women, coping with their different roles and functions, bring their day-to-day concerns and problems to the attention of a Supreme Being. This demonstrates a spirituality that interrelates the physical and the spiritual. While at first glance their spirituality may appear to be an act of powerlessness and fatalism, it seems to me on deeper reflection that, for a woman, religious practice may be an act of 'attunement' with 'forces' bigger than herself. This may come from a realization, or perhaps from a basic intuition, that she exists as part of the bigger world; hence her connectedness and relational character which are affirmed and expressed through her spirituality.

Women's aspirations are central to their spiritual practice. When asked about their aspirations, a common response is 'to continue to breathe' expressed in the Cebuano word, *makaginhawa* (*ginhawa* means 'to breathe'). Beneath the literal meaning of the act of breathing, however, the word implies a longing for well-being or relief from a difficult situation. Elaborating their response, the women talk about relief from economic problems of the family, and include other aspirations, such as seeing their children becoming responsible and succeeding in their studies; having harmonious relationships within the family; seeing their children develop a relationship with God; and making a contribution to the community and family, such as supporting the education of a family member.

Women's spirituality is thus an expression of their *loob*, affirming the *maayong buot* that they want to create within themselves in order to be able to relate to or connect with the greater *kabubut-on*.

Like the ocean wave, the Filipina's self-identity is in continuous motion. It is dynamic; it is both individual and shared. Creating a balance within her *loob* gives her *maayong buot*, which is closest to a feeling of wholeness. As a woman copes with all the expectations and responsibilities laid upon her, she feels the pain of being 'submerged', as in the anguish of the overseas worker who has to leave her family. Paradoxically, she emerges with a *lakas ng loob* because she chooses or has the *buot* to respond to the life needs of her family. Her willingness to accept hardship and sacrifice for the good of the family is important to her. While she carries a sense of mission for the family, however, she is at the same time very vulnerable herself. The trappings of the 'good life' to which she aspires for her family may bring her into situations where she will have to go against her own values and ideals. The ordinary Filipina is faced with the challenge of creatively and effectively concretizing her *kabubut-on* in the kind of development for the country that can bring about 'wholeness' and *ginhawa* not only for herself and her family, but for all families in the Philippines.

6

BELONGING

Yvonne Deutsch, Israel

Dedicated to the memory of Ella Deutsch, with longing

Klara (Kempler) Schlesinger, 1912–

Lia (Schlesinger) Meyer, 1931–

Yvonne Deutsch, 1954–

Liane Rosenthal, 1991–

KLARA (KEMPLER) SCHLESINGER, 1912–

New York, spring 1995

She lies on the couch in her room, in need of rest to appease her tired limbs after having completed the routine chores of housekeeping. I have come from afar to be with her in the big and fascinating city of New York, together with Tomer, my 8-year-old son. She, my grandmother – my mother's mother – Klara (Kempler) Schlesinger. Her face has strong, sharp features, her green eyes shine with the wisdom of a woman of 84. Today, as usual, she is pleasantly dressed, a matching chain around her neck, her hair well cared for. Her image, her reddish colours and her straight chin jutting forward, awaken within me the association of an elderly lioness.

Until I reached the age of 8 she was part of my daily life. Together with my mother, I lived in an extended family house in a pleasant apartment with a large garden, on August 3rd Street, in Timisoara, Romania. Chestnut trees lined the park opposite the house. With Klari I went to the market to taste the cheese which the Romanian farmers brought in huge pots, covered with white cloth; with her I went to the train station to wait for my grandfather, returning from his far-off work as a doctor. She was the one who negotiated with me on a daily basis the endless struggle of what to wear. Over and over again I had to forsake the dress with the fancy velvet to go to kindergarten.

In her youth Klari was light-haired with sparkling green eyes. In her pictures she is dressed sensibly but elegantly; her eyes express frankness, innocence and love of life. She is the only daughter of a Hungarian Jewish family. Her father earned a living trading in wood. The comfortable life suggested by the pictures of her childhood home were disturbed by the First World War, which deprived her of her father for a time. Then, just a few years after returning from the war, when Klara was 14, he died of pneumonia, far away in a sanatorium.

She continued living with her mother, Ilus Kempler, an elegant woman with a taste for a life of pleasure and leisure. At the age of 17, Klara was sent to Vienna to learn German and hat-making. There she met an Egyptian suitor who fell in love with her and wanted to make her his wife. But when he was sent to Sudan, she did not wait for him, but married Dr Laci Schlesinger, a gynaecologist, sixteen years her senior. During the Second World War, her former suitor invited them to emigrate to Egypt and even arranged work for my grandfather in a hospital. More than once she has told me, 'Had things worked out differently you would be an Arab.' For me too – had things worked out differently – I would be bringing up children with a Jewish Arab identity.

At the age of 20, while giving birth to her first daughter, Lia, Klara almost died. Two years later her second daughter, Ditta, was born. Did she suffer from postnatal depression? What emotional developments did she go through? What factors helped her to cope with her transformation from a young woman to a mother? How did this affect the early relationship between mother and daughter? What role did the father play in this complex human experience?

The division of labour in the Schlesinger home was according to conservative role models. My grandfather was severe in his facial expression, he had a developed sense of responsibility and rigid demands for himself and his family. Klari organized the household. During the cold winters she was also responsible for lighting the wood-burning stoves in the large rooms of the house. Her role was to bring to life the warmth of the house. His needs were the first to be satisfied and his opinions were an important factor in any decisions of principle. By the time he returned from work, she would be elegantly dressed once again, after having spent the morning working on household chores together with the servants.

Klari is described by those who know her as a woman with a love for life, humour, people and conversation. I feel for her a great warmth and love. While visiting her I am also aware of a certain harshness which I associate with the strict Austro-Hungarian culture and with classical European education.

Due to Klari's sociable personality and her capacity to welcome visitors, the Schlesinger house was open to friends, relatives, immigrants and refugees who received shelter there. The young guests and the tenants felt wanted and loved. Klari was a kind of intermediary between her spouse and the young generation; she participated in their youthful spirit, enjoyed laughing in their company and was amused by their antics, and by some aspects of life which took place far away from her husband's severe eye. Her ability to run an open household and yet live with her rather stern husband – of whom I have pleasant memories – stems from her natural optimism and her ability to manoeuvre between opposing needs without creating a conflict. She learned to cope with changing factors which were beyond her control. However, avoiding conflict also meant that she did not take a stand in certain spheres. This factor was an essential element in the creation – or perhaps deepening – of a psychological distance which was later translated into physical distance between Klara and her elder daughter, my mother. For more than thirty years now, mother and daughter have lived on different continents. Since the end of those days, long ago, when the Schlesinger home was an open house under Klari's direction, Klari has lived with her younger daughter Ditta and her son-in-law.

In the early 1960s the physical breakup of the family took place as a consequence of a wave of emigrations to Israel, the United States and Germany. The first to leave the house were my mother and I. The last to leave was my grandmother, after my grandfather died of cancer. She was left alone to sell the house and to emigrate to a distant land, the United States. At the age of 56, Klari found herself free to enjoy living for herself – a new experience. She lived in Rome for a number of months, waiting for her entry permit to the United States, learning English and enjoying Rome's classical treasures. Upon reaching the United States she began, for the first time in her life, to work outside the home, taking an office job and showing her command of the new language. She adapted to her new life, far from the land and the culture in which she had spent so many years, and created for herself a new social group. She enjoyed the cultural abundance on offer to her, but at the same time she continued her role of running her daughter's household. Here, too, she is known as a pleasant and sociable woman who has the patience to cope with the demands of others.

During my visit she moved slowly, with the help of a cane, but continued to fulfil certain roles in her daughter's household. I closed myself in with her, in her room, sitting on the floor and looking at old pictures. She immediately raised the topic of war in former Yugoslavia. The world has gone crazy: they haven't learned the lessons of the past wars. The world becomes a more and more dangerous place to live. Her reputation as a dedicated carer of her family, together with her thoughts on the destructive forces within human culture, create within me the image of a goddess who is the defender of life – but an ageing goddess, whose strength to defend life is fading, whose power is weakened in the patriarchal society. Although many women find ways of adjusting to the frameworks of their lives and of playing a meaningful role in community life, women lack real influence on crucial matters. We have no influence on decision-making, on questions of life and death, war and peace. For Klari, her capacity to adjust and her desire for peace rather than conflict within her immediate surroundings undermined her powers of defence. By not taking a stand she fails to use the defending force of the lioness and cedes to those who are considered to be in positions of authority in society.

Thirty-three years ago I was separated from her. I do not have many concrete memories. The eight years during which I was in her care, looked after by her, while she performed many of the day-to-day tasks of motherhood, have left imprinted in me a deep love for her. At the age of 8, I lost the daily touch of Klari. From a concrete figure capable of influencing me deeply, she became an abstract image who comes to life in infrequent letters. The feelings remain after years, even, but there is now a distance due to the lack of a real connection, and I do not succeed in extracting much from her during the few days of my stay in New York. General questions about the life of women are difficult for me to ask, and I concentrate on personal questions. Even the pleasure she derives from her great-grandson saddens me, because she cannot have the same experience of my daughter, cannot get to know the vividness, the humour and strength of Liane, which derive, I believe, from the same potential source of strength within Klari's personality. When I leave, the renewed separation is hard on me; as on similar occasions in the past, I have no idea when – or whether – we will meet again.

LIA (SCHLESINGER) MEYER, 1931–

Tel Aviv, summer 1995

She phones me in the early hours of the morning and in a faint voice she tells me that she is on her way to the hospital. At the age of 64 she has collapsed mentally and physically – the mother of an only daughter, twice divorced, once widowed, speaking six languages, no profession: my mother, Lia (Schlesinger) Meyer.

In January 1996 she turned 65. The years show no signs on her face, but her body is broken. She moves like an old woman, doubled over under the heavy load. She moves with difficulty, with the help of a walker, a woman who abandoned her body to the hands of surgeons to deal with orthopaedic problems. From operation to operation her state deteriorated further and further until she became disabled. How did that pretty child in the picture, with a smiling face and a child's innocence, become a broken woman? What forces have played in the course of her life and on the path she has trodden? In the Austro-Hungarian culture we did not learn to be conscious of the link between body and soul. How do women learn this in a culture that creates alienation between women and their bodies? And how do the surgeons of Western medicine know the bodies of women? What is the influence of the repressed aggressiveness of the surgeons on the bodies of their female patients?

At her birth in 1931, Lia's mother almost died as a result of complications. When Lia was 2 years old, her sister Ditta was born. Klari likes to tell that Ditta was so ugly after she was born that she would cover the baby carriage because of the shame. But she did not stay the ugly duckling for long: Ditta continues the tradition of the Kempler women not only in her looks, but also in her capacity to please and her skill of being able to make the most of her life.

Lia, following the established role of the first-born child, charted the path for her mother and her smaller sister. Her personality blends traits which she inherited from her father, such as seriousness and level-headedness, with certain Kempler characteristics: she has a great sense of humour and love of life. She lives with a deep belief that her fate has been bitter, without any perception of autonomy or direction of her own destiny. Retaining her humour and love of life are thus intermingled with her fight for physical survival and often with serious breakdowns and descents into despair.

Her youth and adolescence coincided with major historical events such as the Second World War, the invasion of Romania by the USSR, liberation from the Nazis and the rise of communism. In addition to anti-Semitism, there were the horrors of war, such as the nightly bomb alerts with shrill sirens and refuge sought in the shelters. Her father was taken to a work camp to work as a doctor and, together with her mother and sister, Lia lived most of the war in Timisoara in that same large apartment building facing the park; a paradise for children, as it was in my childhood. She was one of a group of Jewish youngsters whose sense of community remains highly developed even today, long after her friends dispersed to several different continents. In that part of the world, which had fallen under the control of Romania at the end of the First World War, those young people grew up at an international meeting point of Austro-Hungarian and Romanian cultures, with a clearly felt French influence. They were the second generation of secular Jewish intellectuals, most of them also influenced by the Zionist movement. As well as studying at Jewish schools, they had periods of education in Jewish youth movements, and part of their nostalgia is connected to those experiences. Their political world-view is fed from the horrors of the Second World War and their lives under the communist regime.

Although she was a good student in high school, Lia was forced to leave her studies at the insistence of her father, a respected doctor in the Jewish community and active in the Romanian Zionist movement, in order to study sewing, a profession much in demand in Palestine. My mother was part of the second generation of women to be trained for professional tasks. In her case, her own wishes were disregarded, as were her personal talents and need for professional self-fulfilment. In our family, women work out of the house in order to help in the economic running of the household. The decision as to what Lia should study was made by her father, on the basis of so-called practical considerations. The fact that she hated sewing and was actually talented in learning languages was not relevant for him. But if she is one of a generation of women who learned a profession and left the home to work, she was also, like my generation in another land, meant to fulfil her main function as wife and the mother of children. These are the socially acceptable principles which define the identities of women and which become an oppressing factor of their personal choices.

By the age of 20, Lia had married my father, despite the opposition of both families, because of their youth. Given the conditions under the communist regime in Romania, they could not afford to live in an apartment of their own. The beginning, and also the end, of their married life took place in a room in her parents' home. Two doors led to this room which overlooked a wide boulevard, lined by the park on the other side of the road, with its chestnut trees. In this space, between her parents' room and her sister's room, my mother tried to build her life with that tall and charming young man, with his reddish hair and glasses. At the time he was a rising star in the world of mathematics; he enjoyed music and dancing, had a good sense of humour, was pleasant, loved by all his friends and acquaintances: Emeric Deutsch – my father.

I stare at her picture; she is in her twenties, a good-looking young woman. Despite the sadness emanating from her eyes, the expression of her face does not reveal that which is to be her future. When she was 23, I was born. Her husband was then investing his energies in building a career and falling in love with another woman, like a romantic wind blowing over the stage at the opera. Lia found herself living within her confines and caring for her child. In her group, the others were still free and were able to enjoy to the utmost the rich cultural life in Timisoara, Bucharest, Budapest and Prague. The rest were free to continue their lives, whereas she had to cope with the change of role from young woman to mother.

My birth was accompanied by the emotional upheavals of a young couple prior to their divorce. Within a few years the marriage was untenable. In daily life, dramas of love and separation are not heroic and breathtaking as they are at the opera, my parents' cultural inheritance. Later she would tell me that her dream would be fulfilled with a husband and children. That which she truly and earnestly wanted, she did not succeed in achieving. To me it is not clear if those were her true aspirations or whether her socialization dictated her dreams. In any case, her destiny was not generous, and she who has such a hard time being alone never found her way to a regular family life.

Upon her divorce from my father, she changed overnight from a woman destined to be the wife of a professor to a woman with no profession or formal education, forced to build a life for herself by relying on her own strength, in a society which was not friendly to women and especially not to one-parent families.

My father's departure from the house created a crisis. The words 'hurt', 'pain', 'insult', 'rejection' and 'helplessness' can be related to this part of life. Despite the efforts of the family to salvage the marriage, a conflict was declared between the parties, and I (as the daughter of divorced parents) was between the hammer and the anvil. My mother, whose real talents were not recognized as a potentially useful creative force, began working in an office, while her mother shared the care for her daughter. For five years, until we emigrated to Israel, she continued living in her parents' home, in that house which remained for me a source of unrelenting longing, a family anchor and a place with a sense of belonging – the same house which was for my mother a source of stress from which she wanted to leave. The emigration for her was an escape from the asphyxiation she felt within the framework of the extended family.

However, as in many cases of emigration, it was also accompanied by a drop in her personal and social status. In Romania, her place of birth, she could have advanced in her office work and in time learned a profession and felt sufficiently respected in her social community, the emigration forced her to join the labour force, although her background was of the urban, educated class. An immigrant woman, 30 years of age, the mother of a daughter, with a wide cultural education and command of six languages, began to build a new life. The need to establish herself and to cope in the new society was accompanied by a need to separate herself from time to time from her daughter. She earned a living from unprofessional labour. At the age of 38, her life stabilized with a new marriage: she was no longer forced to take hard physical work, she could once again afford to take office jobs. Ten years later, with the death of her husband, her physical condition began to deteriorate. Before she reached the age of 60, she became disabled. She lives her life as a single woman. The life of one woman, survivor of the Western medical establishment, is a story still to be told.

I am the daughter of that same injured woman who left her home to find a new life, and that very life alienated her from me and broke her.

YVONNE DEUTSCH, 1954–

Temesvar

I was that child, born to a couple who were on the verge of breaking up their relationship. Only the experience of watching the physical expression of conflict between people close to me – the emotional response which rises up in me like a wave, threatening to choke me, and bringing with it the desire to scream or to flee from the situation – only this reminds me of the existential experience into which I was born.

Until the age of 8 I lived in the spacious apartment with its numerous occupants. I grew up with a grandfather and grandmother who loved me, a German childminder who was one of the household, a mother who was trying to rebuild her life, an aunt and her friend, two young working women, pleasant friends out to enjoy life. This was a Jewish, urban, educated family household at a crossroads between cultures, which generated cosmopolitanism on the one hand and emotional closure coupled with political conservatism on the other.

I barely remember the feelings involved in the departure from my family. I presume that the pains of departure from the relatively secure place at the bosom of the extended family were accompanied by appeasing words about the approaching meeting with my grandmother and grandfather (Ella and Geza Deutsch) who had emigrated to Jerusalem a year earlier with their eldest son, Erno Deutsch. The open expression of feelings is not one of the characteristic traits of Central European culture, which emphasizes that which is permitted and that which is not, while repressing all feelings from the surface.

Israel

On 20 February 1962, in the early evening, we landed at the airport in Lod. A child of 8 years old, together with her mother, a woman around 30. A number of days before, we had left the home in which we were both born. The long journey across the continent, in the freezing winter of 1962, was supposed to end in Jerusalem, in the house of my father's parents. But the heavy snow of that year forced us to travel to the sand dunes of Ashqelon, to a distant absorption camp in which the huts were subject to the vagaries of the weather, as though refusing to absorb within them the immigrants who had just arrived.

After one night in a hostel at the airport, we travelled on a clear, cold winter day along the coastal road from Lod to Ashqelon. We travelled together, in silence. 'Look at the groves full of oranges,' said my mother with excitement, trying to contain within her the sorrow and anxiety. At home, in Romania, oranges were exotic; aside from the sensual excitement of their taste, they also represented a longing for the beyond, for another land, a place where dreams are likely to come true. And now, in that faraway country, we travelled alongside groves and discovered the oranges, their colour bright in the light of the cold sun.

The second night in the new land we, my mother and I, spent in a shaky shed. We slept alone on metal beds covered in woollen army blankets, prickly and grey, in the light of a kerosine lamp which shone neither brightly nor warmly. About a kilometre away, at the perimeter of the camp, separating us from the sand and the sea, a foreign man who did not speak our language served as a guard.

The harsh move from life in the bosom of the family to this foreign land, in which I encountered strangers who were presented to me as family, left a deep imprint in my personality and my life. At the age of 8, my roots were severed live. Here, this small land – in the cradle of Jewish culture, Muslim and Christian, the Middle East – became my home. Still eternally searching to deepen my connections to this conflict-stricken environment; the voices of the primal goddesses have not been heard in a long time. And I, still coping with the experience of belonging and strangeness, searching for roots, defy the Zionist myth, walk on the red line, oppose the occupation, dream of a just society not only with respect to the Palestinians but also with respect to women.

Two months after we arrived I was sent to a boarding school while my mother registered for a Hebrew language course. Two months after having been cut from my family I found myself living in the company of children my age whose language I did not speak. The exposure to my mother's struggle forced me to treasure my feelings within me, enclose them and keep going. That same mechanism of closing my feelings within me, keeping my head above water, is to accompany me for many long years, even during times when the usefulness and effectiveness of such a mechanism are doubtful.

The emigration to Israel strengthened the relationship with my grandmother Ella (Stern) Deutsch, my father's mother, to whom I am physically similar. For ten years I used to spend my holidays at her home in Jerusalem. Only after her death did I understand that she was a figure from whom I tried to draw my roots. I was 18 when she died, and although I do not visit her grave often I long for her deeply and my sense of loss is great. She was born in 1900, the eldest daughter of five, the only one who continued her further education and graduated from pharmaceutical school. With marriage and motherhood she stopped working outside the house and abandoned her profession. In my search for models of educated women I try to rest on that of grandmother Ella, although here too the unseen is greater than the apparent. How did she feel when she had to stop working at her profession? What was the meaning of it all to her? What were her opinions on political and social questions? I remember Ella as a serious woman, straight and modest, who was also a mother image for my mother.

At the age of 8 I was thrown into a different cultural environment, taken to a strange land and told it was mine (to the Jew who is in you but not to the woman within you). My acclimatization was apparently swift; I wiped from my consciousness the culture I came from, the rich European cultural life, and exchanged it for secular, Zionist culture, rootless in the Middle East environment. The Vienna dance rooms resounding to Strauss waltzes became hidden centres of longing from my childhood. The paintings of Renoir and Rubens represented for me the bodies of women from the place I came from.

The emigration introduced me as a child to different communities. After a short attempt by my mother to live together with me as a single-parent family and as a breadwinner without adequate financial support, she felt compelled, against my wishes, to send me to boarding school again. An accident in which I broke a door and cut my hand on a splinter of the shattered glass convinced her that I could not be left at home in the afternoons while she was at work. This reality traced my destiny back to school, but this time I chose a girls' boarding school in Jerusalem. In the previous boarding school I was abused by the boys who, together with some girls, formed a group, and in the middle of the night when I was in my bed tried to undress me. When I woke up they burst into laughter. This experience was enough for me to decide to live in a world of girls only. This was to be found in an ultra-orthodox religious boarding school on Bar-Ilan Road in Jerusalem, the place where there is now a battle for the closing of the road on Saturdays. There I stood, thirty years ago, and yelled together with other girls, '*Shabas! Shabas!*' at the passing cars.[1] For two years I lived in the orthodox boarding school with a strict religious system and at home I continued a secular way of life, together with daily prayer. In the boarding school I was the only Ashkenazi girl amongst Mizrahi girls from disadvantaged social backgrounds.

In the summer of 1965 I moved with my mother to live at Givat Shmuel, adjacent to Tel Aviv. There my second home was built. Despite my request to continue the religious education I had become accustomed to, my mother decided on a secular school. And there I found myself transferred directly from the Jerusalem ultra-orthodox environment to the socialist Zionist bosom of the *Shomer HaTzair*.[2] 'If your grandfather only knew that you are being educated in a communist youth movement,' my mother would say, and encouraged me to join. And I thank you, Mother, that you sent me to that group of youth in Givat Shmuel in the mid-1960s. There I learned of the variety of identities in Israel. In the movement we absorbed universal humanistic values. In school we were educated in the myths of the Zionist society. Exposure to the Zionist political culture, nostalgia related to the establishment of the State, pictures of youngsters sitting after a long day around campfires, sipping coffee from the *finjan*, and the feeling of togetherness, were the basis of the Zionist culture which I immediately adopted in my attempts to integrate myself into the new environment. The fact that I had not participated in the establishment of the State made me feel that I had missed out on something. Freeing myself from the myth began at a relatively early age. At the age of 14, reading a love story in the wooden hut of the Shomer HaTzair, I became aware for the first time of the repression of the emotional link between a Palestinian man and a Jewish woman. My short military service in the Gaza Strip and my meeting with Palestinian youth at the university strengthened my coming to terms with the collective denial of the price paid by the Palestinians for the establishment of the State of Israel. It encouraged my growing opposition to the occupation.

This is not the place to describe the conventional life of an Ashkenazi girl in Israel who, upon immigrating with her mother to the 'Promised Land', became the daughter of a chambermaid. I had no family ties to the Ashkenazi Zionist society, no father who participated in that same military ethos which still makes up the political, the social and the economic canon of the State of Israel – no father at all as a balancing factor, to deal daily with the life of his daughter. Letters, postcards from around the world, gifts, ten-day annual visits – these are my childhood memories of my father.

I finished high school in a then-quality boarding school, Hadassim. In the army I served together with my friends from the Shomer HaTzair in the Nahal, which mixes military service with the kibbutz movement. We were known as a group (in Hebrew 'seed') which included leftists and Trotskyists. In the 1973 elections the majority of the group voted, at least in the Gaza Strip, for the left-wing Zionist Socialist party, *Sheli*. After a period of two months, we were transferred from our duties in Gaza in embarrassing circumstances, because of our peace-seeking behaviour as young Israeli soldiers with the Palestinian refugees. Instead of patrolling as required in the refugee camps, we discussed politics: people spoke of the integration between the national struggle and the workers' struggle and perhaps also about the idea of establishing a secular democratic State. We were transferred from Gaza to another camp; the military commander shouted at us. The random authority which enabled our commander's ego to expand beyond reason I can accurately remember; his rank I have forgotten. Although I have no traumatic memories of the army I understood the principle of male military hierarchy before reading antimilitaristic writers such as Cynthia Enloe and Carol Cohen.

At university I joined the Jewish–Arab coalition of the Left. Although I completed my first degree in Hebrew Literature, African History and Social Work, and worked for ten years as a social worker, I never managed to free myself of my social-political activist feelings against the occupation, in favour of peace, and towards the establishment of a political culture of women. In my twenties I joined the movement against the occupation in Israel. In my thirties I participated in creating a women's peace movement. In my forties, as a feminist peace activist, I am concentrating on the establishment of a feminist centre, *Kol Ha-Isha* (The Woman's Voice), in West Jerusalem.

Those years of peace work gave me vast experience in the ways of organizing women politically. In Israel the women's peace movement developed a concept akin to that of the women of Plaza Mayor, and created an international women's protest movement against various manifestations of violence, including war and destruction and violence against women. The movement was born in Jerusalem, created by a small group of Radical Left women, and in Tel Aviv, by feminist women and women from the Left in general. These reactions to the occupation, immediately after the outbreak of the *intifada*, allowed the growth of a women's group which now includes Women in Black groups in countries around the world, culminating in a moving protest vigil opposing violence against women at the recent women's conference in Beijing.

Jerusalem, November 1995

In Jerusalem, the 'city of peace', I walk a thin line. I, who as a child was exposed to a variety of identities, walk the earth with no roots. Here is my home of the past twenty years. Before 1967 we would go up the former minaret of the YMCA or on to the roof of Notre Dame to look at East Jerusalem. Next to the kiosk of Shoshana Shwartz in Abu Tor, the voices of the nearby city echoed, yet still so far. I remember the strange feeling that bustling lives were being led there, although one could not feel the pulse of Palestinian life in Jerusalem. East Jerusalem is no longer the unknown world from which only the voices of those living there cross the barbed-wire fence. I search for connection to this place which has become my home – by way of peace, by coming to terms with the past, the political reality and the roots which emerge from this land.

In the city of peace, occupation encourages conflict. Jerusalem is a microcosm of all the tensions in society. Islands of cultures of peace do exist. In the period of the *intifada*, the women made the most visible contribution by protesting and by creating an alternative of peace. In Israel the protest movement of Women in Black developed, the circles of dialogue between Israeli and Palestinian women grew, more and more Israeli women were exposed to the Palestinian reality of life under occupation, and were compelled to face previous prejudices, racism, and to free themselves from a self-deluding image of a so-called enlightened occupation. At the time of the Gulf War, the women continued their meetings for a time, though the war put an end to this. Most of the women were confronted with the political differences derived from a different personal collective identity: most of the Ashkenazi Jewish women in the peace movement identified with the West, whereas the Palestinians identified with Iraq and its allies. The vigils of Women in Black, demonstrating against the occupation every week at the same place and at the same time, began to fade. With the coming to power of Labour-Meretz and the signing of the Oslo agreement, political peacemaking became establishment work. Those against nationalism and militarism are still in the closet.

The political oppression of women's culture prevents us from participating in and having an influence on matters of principle, of life and death, despite – or rather because of – the fact that our bodies carry life. In the Israeli–Palestinian conflict, even though women have an impressive record of cooperation against the occupation and in favour of the establishment of two states as a means of achieving peace, it is difficult to overcome the division due to the national conflict. Because of the activism of Mizrahiot feminists in Israeli society, there is within the feminist movement a coming-to-terms with other identities, identities which clash in women's lives and which stem from the unjust distribution of the right to self-determination, autonomy and political strength. In this land, the national identity, the religious identity, the ethnic identity and the class identity are all more developed than gender identity.

Although I have Palestinian friends both in Israel and in the occupied territories, our paths seldom cross. Despite the sympathy, understanding and common political world-view, each of us is active within her own community. In Jerusalem, the city of peace, I live within a ghetto – in the margin of politics, within a world of women in the public sphere, with international relationships but without continuity in my relations with Palestinian women. I often think about those women to whom, at different times, I have felt close, in whom I have admired the capacity to act against the occupation while trying to meet Israelis and create a change in Israeli public opinion.

How can I connect myself to this land and create roots here? The political culture in Israel and in the Arab states is unacceptable to me. The emphasis on national and religious identities splits the woman's identity between conflicts which derive from the patriarchal system; the voices of women are not heard. And I, in the alienation from the ruling political culture which exists within me, and in my search for roots, want to build a feminist political culture which has its roots in a global network of women's cultures. This political culture would shape a perception of women as a global collective with historical awareness and collective consciousness. It would represent a combination of experiences which one by one grant us the environment of the cultural feminist reconstruction which we are seeking.

Even if there are differences of opinion among women themselves about the definition of terms such as 'womanly', about what we want to keep within our traditions and what we have to change, surely we can agree that the priorities which configure the military and atomic industries are not women's priorities. We did not participate in these decision-making processes, although many of us do not understand the boomerang effect which is part of cooperation with patriarchal priorities. Many of us do not see the price included in forfeiting self-determination as part of the struggle not to be the 'Other' in a society built on the 'Otherness' of women. We must listen to the voices from the lives of different women as well as those within ourselves, if we are to deepen our own knowledge – voices from the women within us who derive from the different existential historical experiences of women.

Women in patriarchal society cope with the inner contradiction between the search for historical reconstruction of humanist cultures and the urge to belong to a community. The urge to belong creates the cooperation of women with the patriarchal culture. The internalization of oppression causes women to cooperate with priorities which involve the evil within us. A friend from New York told me once that the greatest oppression she has experienced was by women. We must be conscious of the personal and political traps in the process of personal development and the building of solidarity which crosses borders among women. Our love creates within us the commitment to struggle to ensure the continued existence of humankind.

I am tired. The distance between the vision and reality makes me anxious. The fact that, at the age of 41, I still pay economically for the urge to be engaged in feminist political work is frustrating. But the potential for the destruction of humankind which expropriates my power as a mother to defend her children and ensure their personal security is the largest sore among all the sores which I have known during my life. The fact that their destiny is determined by crazed generals and men desperate for power, strength and riches, makes me mad.

LIANE ROSENTHAL, 1991–

Jerusalem. Nahlaot

I look at my daughter, Liane Rosenthal, aged 4. Light-red hair, tall. She has a strong presence, speaks in loud tones and demands that which is her due – strong, social, seeking adventure, warm, domineering, stubborn. Her body movements express confidence. She is not afraid to jump from high places. At an early age she insisted on learning how to use roller-skates. She has within her a mixture of assertiveness, curiosity, positiveness, knowledge and the charm of a 4-year-old, together with a demanding nature and a need for control. She has a strong thirst for warmth, for connection, and at times a fine thread of embarrassment hidden in her reddish personality. She has the strength to persevere and a desire to learn and succeed. She has learned to express herself directly and clearly, and she mixes words with strong facial expressions. She walks in the world as though it belongs to her. At the moment I feel that if she wants something she will probably be able to accomplish it. What are her possibilities and what will be her future? What incidents will be the potential breaking points which crack open the protected world in which she lives today? In contrast to other children of the area, my children, Liane and Tomer, were born into a privileged social class, with Ashkenazi parents who made the choice to become activists in the heart of the Israeli–Palestinian conflict, to oppose the establishment and therefore to position themselves in the margins of the ruling ideology.

I have no doubt that Liane will have the strength to fulfil her aspirations, although the world in which she and Tomer were born is not secure, and this knowledge pains me.

REFLECTIONS

The Encounter with the authors of this book brought me back to Brussels, a city I had visited in the past. As a 'Woman in Black' in Israel I have had the opportunity to participate in a number of these women's conferences. As an inn for resting and changing horses after a long ride, they help me to renew my energy before continuing. But this time it was different. Our stay together, the dialogue linking our experiences of life as women from different cultures – while within ourselves we carry historical continuity of cultures as women – created within me an inner feeling of peace. Like a sponge whose spores crave water as a life source, I absorbed within me this spiritual experience.

I am in the midst of a period of intensive political activity. Immediately after the outbreak of the *intifada*, I joined the activities of Women in Black and took a central role in the establishment of the women's peace movement in Israel. With the co-option of the peace movement into the establishment, I turned to the forming of a feminist centre in Jerusalem, *Kol Ha-Isha*. These recent years feel like a marathon effort. Stemming from a deep feeling of responsibility and of my duty to take a stand in the Israeli–Palestinian conflict, from a deep feeling of solidarity towards women and a desire to change the status of women in society, and from an existential anxiety born of the destructiveness ingrained in the patriarchal system, I dedicate myself to feminist political activities. The grind of work, my exposure to women of conflicting identities, the need for integration between my public work and my private life – all leave me with little time for those inner feelings of peace.

But our joint experience in Brussels has enabled me to renew my search for personal and cultural roots. At the age of 8 I was uprooted from the place of my birth and from my family. The reality of my family's emigrations forces me to cross time and space in order to reconnect the roots of my identity. In this inner journey of consciousness there is a renewed attempt to come to terms with the feelings of loss which escort me through life. The return from the political to the personal, my participation in this book, forced me to slow down the rhythm of my activities and allow the contents of the past to re-emerge in my mind. This slowing down is also necessary to achieve some balance between my mind and my body, and between my role as a mother and my feminist work. But the stream of activity is like a wave which overcomes me. Is it by chance that I began to write this story between *Rosh Hashana* and the Day of Atonement, a time of re-examination in Judaism?

NOTES

1 *Shabas* is a Yiddish word for 'Saturday' which in Hebrew is *shabbāth* (Shabbat). According to Jewish religious belief, Shabbat is sacred and a Jew is not to defile it by driving between sunset on Friday and sunset on Saturday.
2 *Shomer HaTzair* is a feminist youth movement with a socialist orientation, which is active in kibbutzim, cities and villages in Israel and abroad.

7

NO ISOLATION ANY MORE

Amal Krieshe, Palestine

Mas'uda, 1887–1987

Nabiha, 1910–

Amal, 1957–

Itisal, 1973–

MAS'UDA, 1887–1987

'Her independence was a challenge to the habits of the family'

My grandmother was called Mas'uda, which means 'lucky woman'. She lived from 1887 to 1987 in the village of Dinabeh in the north of the West Bank.

It looks as if during the hundred years of my grandmother's lifetime more changes have taken place in the life of the Palestinian people than in the preceding centuries. The Arab region suffered 400 years of darkness under the yoke of the Ottoman empire, and this darkness affected the life of every individual, particularly the lives of women. In Palestine, every village was a self-sufficient economic unit, unconnected to other regions. The nature of economic life was largely agricultural and agriculture was directed at self-sufficiency. The main characteristic of the economy was simple production, with two objectives: own consumption and exchange. Most of the income from the latter went to the big landowners and the tax collectors. Social relations reflected the feudal system. The oppressive local customs and the social system contributed to reinforcing the inferior position of women.

Socially, the Palestinian population was divided between the city and the countryside. Culturally, religion was the distinguishing factor in education up to the middle of the nineteenth century: education was restricted to Koran schools in which male pupils learned the principles of reading, writing, mathematics and the Koran. The wealthy sent their male children to be educated in the universities of Turkey (Istanbul) or the Azhar University in Cairo. However, the large majority of the population was uneducated: at the time of the First World War more than 80 per cent of the Palestinian population was illiterate.

Between 1882 and 1914, Palestinians witnessed waves of Jewish immigration. When, after the First World War, Palestine was mandated to Britain, the Balfour Declaration of 1917 gave Jews the right to reside in 'A national homeland for the Jews', resulting in ever-increasing Jewish immigration into Palestine. During the first Israeli war in 1948, the Palestinians fled their villages and cities in terror, fearing a repeat of the massacre of Dei Asian in which pregnant women, children and others were brutally murdered by the Jews. Hundreds of thousands of Palestinian families were dispersed through the neighbouring Arab countries to live in refugee camps.

In 1949, the General Committee of the League of Nations decided to divide Palestine into an Arab and a Jewish state. After the defeat of the Arab Nations in 1967, the West Bank and the Gaza Strip fell under Israeli occupation.

The Palestinian people reacted to these political developments with revolution and the organization of popular uprisings in 1921, 1929 and 1933. In 1963 a general strike was held which lasted for six months.

The economic structure of the old Palestine was largely destroyed as a result of these developments, especially through the loss of land which was considered central to production. Economic changes were accompanied by social changes, the most obvious of which was the increased interest of Palestinians in education, and changes in habits, tradition and customs related to women's education. These changes created continual tensions between my grandmother and her daughter. Traditionally women had no influence either on public life or on their personal lives; even the women's societies, formed in the first half of the century and made up of women from the upper classes, only supported the national struggle and concentrated their efforts on the cities. At the beginning of the 1970s women's committees were created by leading women from the middle class. My grandmother, who had never taken part in women's movements, saw her granddaughter devoting her time and energy to these women's committees. As a result of the distinguished role of women within the political struggle, the challenges from the occupation itself and the growth of fundamentalism, renewed interest was given to the position of women in law and politics; this changed women's views of themselves and society's view of women away from the patriarchal system.

'Lucky woman' Mas'uda stood out because of her blue eyes and her fair skin. In 1903 she married her cousin when she was 16, in an arranged marriage. She never talked about her views on this issue, or her feelings about it, because she considered it quite normal and not open to discussion. Cousins are important in our culture. Should a man decide to marry his daughter to his cousin even after she has already mounted her horse on her way to marry another man, he can stop the wedding.

My grandmother was very beautiful, healthy and able to have children. However, she could not read or write, because of the lack of schools in her region. There were some Koran schools in the area, but these were only for boys.

My grandfather was a peasant who worked on his land. Out of respect for her

103

beauty, he did not ask my grandmother to help him on the land, but hired others instead. She was in charge of organizing the domestic economy – the production of cheese, yoghurt and olives – and lived with the extended family of her husband. She dried vegetables in summer, baked bread and did the cooking, in addition to the other chores of housekeeping. Her social relations were restricted to what is called 'the circle of the oven'. In traditional Palestinian villages the oven was shared by a number of families because of the effort needed to clean the oven, to prepare it and to find enough dung to stoke it; moreover, one oven produces far more than the needs of one family. Thus, the oven provided my grandmother with the only chance to talk to other women, especially since she did not work on the land (the other source of social contact for women). Naturally, other forms of entertainment were limited at that time: there was no television or radio in the village.

My grandmother had six children, two of whom died in infancy: there were no medical clinics, and medicine was mainly practised using herbs and following local customs. My grandmother was left with just two boys and two girls, at a time when the main task of women was seen as having children, and particularly male children who could work on the land and provide for their parents in their old age (there was no such thing as social security).

My grandfather was drafted into the Turkish army and took part in the popular uprising against the English mandate. He died as a young man and left a beautiful widow, still young, with four children. With his death, her sexual life ended, as tradition does not permit a widow to remarry: she has to devote her life to her children. My grandmother used to wear a robe over her clothes, as well as a head cover.

My grandmother sent her sons to the city to study. The first son finished secondary school and became involved in political activities. The second one graduated in Agricultural Engineering from the University of Damascus. My grandmother went to visit her son in Syria: she returned to an increased social standing, not only within the family but also in the village, as she was the only woman at that time who had left her village and visited another country. She travelled alone and when she came back she described all the streets, the women and the mosques of Damascus. Even the furniture of her house set my grandmother apart: she had rare souvenirs from Syria and from other countries which her sons had visited.

The political activities of her second son further reinforced her position as an unusual woman in the life of the village; her son became an official for the United Nations programme for Palestinian refugees (UNWRAP) and her house became a visiting place for women and men of the village. This gave her the self-confidence to talk about politics. She continuously criticized the Jordanian system which imprisoned her son for ten years. (He escaped in 1957, but did not return to the West Bank – he died in Syria.)

My grandmother insisted on growing orange trees on the land which was left in her name, despite the absence of her sons. She wanted to live alone with her things in her house. Her independence was a challenge to the habits of the family, even generating hatred between her and her son-in-law (her daughter's husband, my father). Her relations with the women of the village were also tense because she was an unusual woman who had achieved so much that other women could not achieve. At that time there were no women's movements in the rural areas, depriving her of a chance to be a role model to inspire other women.

NABIHA, 1910–

'A chain of challenges'

My mother is called Nabiha, which means 'clever one'. She grew up in a house where education was encouraged for males and where women did not take a direct part in the agricultural activities. She married late, at 18, although her exceptional beauty had brought her suitors ever since she was 12. Her brothers, however, refused to marry her off at an early age. They also refused to let her go to school (although there was a school in a nearby village), fearing that she would learn to write love letters to boys and would have chance to make relationships on her way to school. Her marriage was arranged, like that of her mother, and she also married a close relative. She was engaged for a year. In that time, the groom visited her very infrequently and they were always under the close supervision of her family; she never talked to her prospective husband alone. She thought of this as completely normal. The reason for these social habits is the belief that the body of the woman causes temptation and desire; men are prone to succumb to this desire (indeed are unable to resist it), leading to inevitable corruption. The problem can be solved by the seclusion of women in the house and by clothes that cover women's bodies and hair completely to protect men from temptation.

My father worked for the British police and then for the Jordanian police. He was often away from home for long periods of time. This led to natural contraception, so my mother only had six children. During the first years of her marriage, my mother lived with her husband's family, and my mother and grandmother quarrelled constantly. The effects of poverty, oppression and seclusion inside the house are often vented on the daughter-in-law, and quite hateful relations often mark the lives of women in the villages. My mother did not leave the house except for feast days, mourning and weddings, and even then she would be accompanied by one of the members of the family.

When my father was stationed in Jerusalem she went to live with him, returning to the village in the holidays. This gave her some social standing, which influenced her own way of dressing and led her to dress her children in the modern way. When my father decided to build a house in the village he asked my mother to supervise the construction because his brothers had all emigrated to the Gulf countries and his father spent most of his time working on the land. My mother's new responsibility inflamed the strife between her and my grandmother, and even began to involve my grandfather, who considered it contrary to custom and tradition and a snub to his authority. Nevertheless, my mother won the struggle: she supervised the workers and even the manager, and she decided on the layout of the rooms. She was very pleased with her achievement – her self-confidence grew, and she was never the same again.

My mother played an essential role in the education of her sons and daughters, which caused a conflict between her and my father after he came back to live in the village. She was used to her independence in the building of the house, the spending of money, the raising of the children and supervision of their education. We went to her rather than to Father with all our questions and problems, because of the strength of her character. This made her confident enough to discuss current affairs, even in front of men. She made a conscious decision to restrict relations with my grandparents for her own and her children's benefit.

However, the conflict blew up again when she decided to go on a literacy course and her social circle and activities widened as a result. She now had a life away from her domestic activities, which further increased the alienation between her and my father. Still, she did not consider divorce, which is a generally unacceptable (and misunderstood) concept in our society. My father eventually gave in to her – he continued to criticize her, but this did not influence her activities.

The generations of my grandmother and my mother witnessed the Arab Awakening, especially in the field of women's education, work and the issue of the veil. However, the reformist thinkers dealt with women *en bloc* as a 'reform issue', seeing marginalized women as an entity and considering them as instruments for the social and economic development of society. Most of the reformers were men who believed in a gender division of society, including the acknowledgement of the essential differences between men and women. There was no space for individualization, and no social support system which could empower the capacities of women. Neither my grandmother nor my mother took part in any women's organizations.

AMAL, 1957–

'A struggle for equality and social and political peace'

I was born in 1957. My name (Amal) means 'hopes'. I lived through the war of 1967 as a child. In my infancy, my uncle had bought the first television in the village: I saw an Egyptian military show, but I did not understand what it meant until my family began hiding in the house of my grandmother, until I heard the whistling of bullets and saw the Israeli soldiers spit at us after they had locked the people of the village in a square.

The issue of girls' education was not discussed in my family. Although many suitors came for me when I was 12 (I had inherited my grandmother's blue eyes and her fair skin, which are considered valuable assets in the marriage market), my mother became distressed by each such visit, and insisted on educating me instead. My adolescence completely confused me, as I did not understand the process of transformation into a woman. Issues related to sex are not open to discussion within the family, and even if the school curriculum includes them the teacher does not mention them.

My mother's independence from my father freed me from wearing the head cover. My grandfather used to scream at me because my school uniform did not cover my knees and because my hair was not covered. My mother took this as interference with me personally and she challenged him accordingly. My mother did not talk to me at all about sexual matters, but she constantly gave me subconscious messages about the limitations of relations between men and women. I did not have any relations with men during my adolescence. When my brother, who was six years older, tried to make me cover my hair, my mother took a neutral position between us: he was her son and not her father-in-law. This meant a long and lonely struggle for me, but eventually I won my battle.

I found much self-realization through school, extracurricular activities, and in particular the theatre. Another struggle ensued when I appeared not only in front of all the schools, but also in front of the men of education, administration and the press in the nearby town of Tulkarem in the play *King Oedipus*. Nobody at home supported me – my mother considered it a scandal. I challenged them again, and once again I succeeded. After the showing of the play on television, my mother felt very proud, but she advised me not to tell my father of this incident.

My relationship with my father was excellent and he treated me with love and tenderness until I was summoned to the secret police. After that, our relationship deteriorated; he considered me a major problem in the family, fearing that I would not marry as a result of this incident (he had dreamt of marrying me to the son of his brother). When I finished secondary school, my father insisted that I go to the girls' college and not to university. Another challenge . . . This time I went on hunger strike; my mother and my brothers supported me and so I set off to study abroad, in Jordan. At university I took part in political and trade unionist activities and became involved in social issues through the Illiteracy Elimination Programme and the Women's Union of Jordan. My family's respect for me, my personality and my role grew, and I became a source of pride for my father and mother.

At the university I had my first love affair, which ended as the result of a major conflict between us. Because of my involvement in the Women's Movement in the Palestinian refugee camps, my life became very busy. This was difficult for my boyfriend, because I could not devote all my time to him (which was what he wanted). Then I started to witness women's suffering, as workers, wives and students. I found myself full of anger and decided to finish my first relationship, because I did not want to be another slave to men. I later became involved in the cultural activities of the Jordanian Writers' League. In all Arab novels, stories and lectures, women are pictured as weak human beings, only here for the enjoyment of men. This renewed my feelings of anger and spurred me on to political activity. Together with four other women, I led an uprising at the University of Jordan, in which 5000 students participated. The secret police were furious; but, rather than arresting us – as they did the male teachers who were involved – they decided to dismiss us for one year.

This experience greatly increased my self-confidence and assertiveness. I managed to convince my parents to accept this. During this year I joined the trade union, in which I was one of just four women members. In the context of my village this was quite difficult: the women considered my behaviour abnormal, in spite of the fact that they are exploited in factories, have no labour rights and are sexually harassed. This made me start thinking about organizing women, not directly in trade unions but in working women's committees, to empower them and to teach them how to be active in male-oriented organiza-tions. During this process we received great support and solidarity from the political parties – although we later realized that they did this in order to increase their influence and power and not for the sake of justice and equality among men and women. For me, the important thing was to realize both my gender identity and my national identity, and to find ways of using these identities within my political party – to gain influence for my interests as a woman in their policies and programmes, to promote my goals of building justice and to present my perceptions about the role of the women's movements in building and shaping the socio-economic development of our society, especially after the launching of the peace process in the Middle East.

This is a complex process – it demands of me that I communicate with myself so as to realize my identity, and my role and rights as a female human being. I have dedicated myself to pulling the Women's Movement out of its patriarchal relationship with the political movement. Of all the battles in my life, this has been the longest, because new concepts have arisen around the meaning of soli-darity – *al-tadamon*[1] – concerning women's issues, empowerment in the sense of taking decisions in private and public life, analysing everything, including Islamicist principles, human rights and the traditional culture. All this affected my marriage, which was already unusual according to the social expectations of my parents and my political party.

My decision to marry had caused another major battle with my father and mother. When I announced my relationship with my husband-to-be and they found out that he was a politician and was not rich, they refused to accept him. But they knew I would insist and, with the support of my brother, I succeeded in marrying him. I gave birth to three children, and the daughter from my husband's previous marriage also lives with us. In November 1994, I was involved in a terrible car accident in which I lost my eldest son, Anan. The loss of my son was the deepest pain I have ever experienced in my life. At first, I could not imagine that I would ever be able to return to normal life. I felt that my life was finished, like someone drowning in the ocean. But after a while the love and support of my husband, family and the local community helped my soul to revive, and encouraged me to start living again, knowing that human life is the most valuable thing in the world.

My awareness of my gender identity created contradictions and struggles within my married life, for although my husband was used to helping with the children and assisting in the household, and although he supported my role and work in the Women's Movement and the political arena, there is another view of women, deep within him, which sometimes surfaces and upsets our relationship. In my view, this shows the need for the education of men, to help them understand the new type of woman – an individual in her own right and no longer a victim of the historical patriarchal relations in which women have been submitted to the benefit or vision of men.[2]

ITISAL, 1973–

'Seeking her social identity'

My husband's daughter was born in 1973. Her name is Itisal, which means 'connection'. Her mother gave her this name after her parents were divorced, because she hoped the child would reconnect her with her estranged husband. Itisal lived with her mother until she was 10 years old, and when her mother remarried she came to live with us. Itisal's generation no longer has to deal with the problems of dress, education or, to a certain extent, sexual relations or work. But the pressure of society still prevents them from taking decisions about their personal life, because the authority of the father is still sacred within the family. Itisal is a very daring person, creative in thought, thoughtful in discussion and very aware of women's issues. She does not believe that changes in society need time – which causes some problems in the relationship with her father. Itisal finished secondary school and studied Business and Administration; she now works in Jerusalem and is developing her skills further through various courses. Together with sixteen other women, she took part in a course to support women who are victims of rape and sexual harassment. This also helped to raise her gender-awareness and to increase her sympathy and solidarity with other women.

Her social relations are mainly through a '*dabke*' club (Palestinian traditional dance), an aerobics club and the youth club. She has come to the conclusion that she cannot trust men, and feels disgusted by the idea that they look at her as a body only. Nevertheless, she has male friends, she has many male penpals from other countries, and she has taken part (as one of very few women) in meetings between youngsters from Palestine and Israel in England. This has given her more self-confidence, which is reflected in her relationship with her male manager at work. Since starting to work there, she has demanded her rights according to the Labour Law.

The loss of her brother Anan brought her closer to her sister and her younger brother, and also to me. Sometimes she does not know what she wants and needs, which makes her angry. This influences her relationships with males; building up a love relationship is difficult.

Itisal belongs to the generation which will live through the period of political independence with all its horizons of economic, social and cultural development.

The four generations each followed different roads or ways of life. Each road had (or has) its own shape, junctions and shadows. But there are certain common lines which sometimes connect the different roads: the patriarchal society, the inferiority of women as approved of in family law, which is based on the Shari'a law (Islamic law), the fear of the future due to the political situation, and the process of realizing our identity. The vehicle which we four women have been – and still are – driving is fuelled with self-confidence, love and dignity, which all help us to follow our path of life with new hopes and aspirations, and which will bring us finally to a new stage where we can be visible and not isolated any more.

NOTES

1 *Al-tadamon* is the Arabic word for 'solidarity', which is a unity of feelings and helps you to love and to be loved by other people who share the same interests, visions and objectives in life.
2 Through my experience in the struggle of the Women's Movement, together with my female colleagues I developed the concept of gender according to the Palestinian reality. We influenced the political movement to address the issues of women in their electoral programmes on 20 January 1996, and in doing so we started a new phase of lobbying to advocate women's rights as part of human rights.

8

ADVOCATING ISLAMIC RIGHTS

Safia Mohamed Safwat, Sudan

Amena Yussuf Omara, 1899–1962 (?)

Galila Samara, 1917(?)–

Safia Mohamed Safwat, 1944–

AMENA YUSSUF OMARA

Her father's daughter

Amena Yussuf Omara (my grandmother, my mother's mother-in-law) was Sudanese, of ancient genealogy. She was the only daughter of Yussuf Omara, a rich slave-trader who was a *sirr al tujjar*, or President of the Chamber of Commerce. Yussuf Omara was a descendant of an ancient Funj clan; his estates bear the Funj royal seal, a hallmark of sixteenth- and seventeenth-century sultans' land charts. Yussuf Omara was too rich to sell grain during famine and disaster: he piled grain in front of his house for the needy to come and help themselves. This was an old tradition, which came down from the time of the Pharaohs, by which the rich heads of tribes and Sufist *tariqas* would provide for their followers during times of recession, natural disaster and famine. This act of his is still the theme of folksongs on paternal generosity. The praises of his daughter (Amena Yussuf Omara) are also sung, as the epitome of nobility, dignity and charity; she inherited her father's estates, wealth and slaves. (She kept her slaves until she died: although slavery was officially abolished in the Sudan in 1933, they continued to live with her, as they had nowhere to go.)

Her husband, Mahmoud Hussein Safwat, was Egyptian of Ottoman origin; his family had come to the Sudan with the 1821 invasion by Egypt, on behalf of the Khedive in Constantinople. The Egyptian/Ottoman administration was interrupted by a brief period of Sudanese independence under the Mahdist revolt against the Turko-Egyptian rule which overtook the four-centuries-old Funj sultanate. Amena Yussuf Omara's husband later returned and worked with the Anglo-Egyptian administration as Senior Registrar at the High Court of Justice. The civil service in the Sudan has old connections with slavery: government officials, like the military, were the slaves, ex-slaves or foster children of the ruler, the head of the tribe (sheikh) of the Tariga. In her heart of hearts, Amena was not proud of her husband's profession. Although he came from a relatively developed society and was educated, he was a government servant. Exactly how her marriage came about was not known; what was clear was that she felt superior to him from the very start. In Amena's eyes, her husband remained her inferior.

He seemed to have earned additional scorn from her when he had a secret relationship with one of her domestics, living with her as his concubine. It was not until the woman had borne four of his sons that Amena found out: she never forgave him for it and refused to manumit the slave woman or her children. Long after the boys became men, Amena could not find it in her heart to free the slave woman, so that the sons would be freed.[1] The boys and their mother remained as live-in domestic slaves to Amena until her death, even though they were educated and acquired economic independence with good government jobs.

Amena Yussuf Omara longed for a traditional wedding for her eldest son, Mohammed Safwat, with a Sudanese bride who would take on the role of a traditional daughter-in-law to Amena. Sudanese weddings are colourful and prolonged ceremonies. Those of the large old families are especially elaborate and ritualistic, involving exotic cosmetic preparations; stag parties; the rituals of carrying the *al-Mahar* (dowry) and the presents to the bride's house, in processions of relatives and slaves; music and singing; and endless feasting and exchanging of gifts. In the larger families, such occasions are attended by relatives who come from all parts of the country, and particularly from the home town of the family. Wad Medani, the home town of Amena Yussuf Omara, is an ancient town and the capital of the Gesira; it lies on the Blue Nile, where Amena had her extensive estates. Such weddings include conspicuous displays of wealth, emotion and family presence, support and reciprocity – these are intended to express the difference between the important families and those of little or no consequence. Amena Yussuf Omara was more than willing to pay for her son's wedding, just to see him in the traditional costume with a Sudanese wife, preferably a relative. But instead Mohammed Safwat married a non-Sudanese woman, my mother. To Amena, this was a grave injustice on the part of her son, and the cause of deep resentment against my mother.

GALILA SAMARA

Bringing up her children in their own country

Galila Samara, my mother, came to the Sudan with her family. Her father, Mahmoud Fahmy Samara, was an administrator who came to the Anglo-Egyptian Sudan as a representative of the Egyptian government. Galila's father was well-educated, according to the standards of the time, and occupied a respectable position in the Anglo-Egyptian administration until he returned to his native Egypt. Galila was the youngest daughter of a family of six girls and six boys. She was educated in the Sudan, learned English, and was talented in painting, drawing and needlework – the skills which girls from big families were expected to master. Galila's grandfather was an officer in the Egyptian army, her mother his eldest daughter. As an army officer he was socially liberal and had sent his daughters to school. They spoke French, which was the second (or intellectual) language of the middle classes and a status symbol by which they strove to emulate the Egyptians, the standard-setting aristocracy of the time.

Galila wrote poetry. This seems to have run in the female family line; her mother also wrote poetry. Like most women, however, both Galila and her mother kept their poetry to themselves and never published. Galila wrote traditional Sudanese poetry and kept diaries (which later antagonized her husband). When she was young her beauty was renowned; even as she grows older, she is considered remarkably good-looking for her age. At the age of about 15, Galila married Mohammed Safwat, son of Amena Yussuf Omara, and the eldest child of a large family.

The extraordinary background of her mother-in-law, her own experiences and character, and other ingredients of family life all combined to make Galila's marriage and life in the Sudan with her husband's family extremely difficult. Coming from a family which had great respect for family life and family values, she would not betray her upbringing, but continued to struggle for her marriage and her family against tremendous odds. Galila's husband was regularly transferred to the provinces, and Galila would follow him: those times away from his family and their influence were perhaps the happiest in her life. But Amena Yussuf Omara was a powerful woman who was used to exercising control, both as the head of her large extended family and as a rich patron of men and women alike. She had strong opinions about her sons and daughters, making it difficult for Galila and Mohammed Safwat to have an independent life of their own within the context of the extended family (mother-in-law, grandmother-in-law, and sisters and brother of Mohammed Safwat). There was an additional problem, resulting from an initial understanding (or misunderstanding) that Galila's two older brothers were to take part in exchange/arranged marriages to two of her older sisters-in-law. When this 'arrangement' did not materialize, Amena Yussuf Omara held it against Galila.

Galila's marriage began to falter from the start. The social and familial pressures were relentless. The cultural differences, and the fact that Galila's mother-in-law was widowed early (her husband Mahmoud Hussein Safwat died at the age of 45, following an operation in Cairo), added further complications to an already difficult relationship. Amena Yussuf Omara resented the fact that Galila was Egyptian, a foreigner. During periods of particular tension, and at times during her frequent pregnancies, Galila went back to stay with her parents in Cairo. She went through more than nine pregnancies, including two sets of twins and one set of triplets. Her child-bearing started at an early age: she was barely 16 when her first son was stillborn, as a result of chronic malaria Galila had contracted in southern Sudan during her husband's provincial missions.

Repeated visits to Cairo to give birth, or because of tension with her in-laws, meant leaving her husband alone under his mother's strong and persistent influence. Galila's absences gave Amena increased leverage to put pressure on her son to marry one of her relatives. She succeeded more than once, only to be repeatedly disappointed because the marriages did not last long. Mohammed Safwat was deeply in love with Galila and remained so for a long time. This changed only slowly, partly because of the impossible circumstances and partly as a result of Galila's frequent and extended visits to her family in Cairo.

Galila had a strong belief in education for girls, and insisted upon it for her own daughters, making sure that we were educated even when my father was transferred to provinces where there were no girls' primary schools. She tried to teach her sisters-in-law to read and write, which drew strong disapproval from their mother; Amena regarded education as unbecoming for girls from good families. According to tradition at the time, well-off families had people to do things for them: they did not need to read or write. Education was for people who wanted to work in the government, which was not regarded as employment that a well-to-do man or woman would seek.

Amena maintained an attitude of superiority over Galila all her life, and never forgave her or her family for depriving her son of the traditional marriage ceremony. It was partly her constant complaints that ensured the ultimate failure of the marriage. After a series of separations, the final divorce eventually took place. Galila had four children, the eldest of whom was in her early teens. In a traditional society it was difficult for Galila to be a divorcee living on her own with very young children and without an adult male. Her family tried to insist that she return to Cairo and live at her parents' house, but Galila decided against that. She rightly felt that her children would have a better social future in their own country and among their own people. Her family subsequently disowned her: only her mother kept in touch with Galila in secret, sending presents from time to time, on religious and other occasions, via travellers.

In the absence of any family support and as a foreigner in a country that may not have accepted her fully, Galila was an extremely brave woman in deciding to stay and bring up her children in their own country and culture. She had to draw on her own social, emotional and financial resources to bring up her children on her own. Since she was very good with her hands and could crochet, embroider and sew, she took up dressmaking. She was thus able to bring up her four children and help them through their secondary education and university. The family tradition of higher education for girls was continued and passed on by Galila. After the death of her mother-in-law, Galila enjoyed better relations with her sisters-in-law. They followed her example, and most of their daughters – and other girls in her ex-husband's family – went to university and held senior professional and government positions. The death of Amena liberated her daughters and made these decisions much easier to take.

Galila is now in her seventies; she has seventeen grandchildren and two great-grandchildren. Her four children live and work in various Arab and European countries: they either had to go to other countries for education and decided to stay on, or were forced to live abroad because of the situation in the Sudan, where a political fundamentalist regime has been in power since 1989. Galila moves around her family, living with each of her children in turn.

SAFIA MOHAMED SAFWAT

A lonely child in a wild and beautiful country

I was born of an Egyptian mother, Galila, and a Sudanese father, Mohammed Safwat. I am the youngest of four children, with two sisters and one brother. When my parents separated, when I was 9 years old, my brother and sisters and I lived with my mother in Khartoum, the capital of Sudan. Before that, we had lived in various parts of the western and southern regions of Sudan, as my father, a senior police officer with legal training, worked for the Civil Aviation Authority and was frequently transferred from place to place. I had to be taught at home, as there were no schools for girls at that time except Church schools for Christians. My older brother and sisters went to schools in Khartoum and lived with some relatives there. My mother started teaching me at the age of 4, and continued until I was 9. My mother took my education very seriously and I had proper lessons in various subjects every day. But I was all alone – without my brother and sisters around I felt extremely lonely. When my mother realized this, she decided to provide me with the company of other children: she simply opened a school at home, where she taught the other children in the neighbour-hood free of charge.

That part of my childhood, in a wild and beautiful area of Sudan, had a great influence on my personality and on my mental and emotional make-up: so, too, did the separation of my parents when I was young, and my own separation from my brother and sisters for long periods of time. As a child this seemed completely unjust.

When we moved to Khartoum, I went first to a convent school, but later left to attend the Egyptian Coptic College. In order to secure a place there, I had to sit external exams, and although I was younger than the others my grades were higher than the class I was supposed to be in according to my age. My early experiences at school were far from pleasant – I was not used to studying outside my own home, nor to the different routines which were imposed upon me. The sudden company of so many other pupils was something I found very hard to cope with after the long periods of loneliness which had characterized the larger part of my childhood.

With the help of a friend whom I met on the first day at school, I was eventually able to overcome these feelings and get on with my studies. The fact that I excelled at school probably added to my loneliness: it set me apart from the other children and made me grow aloof, especially during adolescence. I was drawn towards reading at a very young age, and read all the books I could find, in the school library and at home, and especially the books that belonged to my eldest sister and my brother. These were mainly books on Marxism, on arts, literature and classics. My brother and sister were both keen readers and were gifted in poetry and creative writing. I grew to love painting and writing and benefited from an extremely sharp memory which helped me remember my lessons very easily.

Between the ages of 9 and 13, I spent a great deal of time with my paternal grandmother who loved and cared for me in spite of her apparent rejection of my mother for being a 'foreigner'. It was during one of these periods in my grand-mother's house that I became aware of class division and the slavery system. This realization came about through a painful experience. I had noticed that a number of women servants used to serve my grandmother, all of whom uncovered their heads and walked barefoot in her presence. When she went out she would always wear a freshly washed, white *toube* (Sudanese sari), and when she came back the *toube* would be taken away by her servants and washed immediately.

When I expressed my surprise at my grandmother's need to have so many women serving her, and asked why they had to uncover their heads and walk barefoot in her presence, my grandmother became furious and shouted at me: 'What business have you got with that? Do you want to spoil my slaves?' I was astonished by my grandmother's furious outburst, as I had never experienced such anger before. She punished me further by refusing to speak to me for the rest of that day, leaving me to cry until my father returned later in the evening.

My father told me that I was not wrong to ask these questions, but explained to me that my grandmother was not unkind to her slaves and that, even if she asked them to leave, they would probably prefer to stay as they would be better off remaining part of the household. He also explained that my grandmother had inherited these women from her father and that slavery, even though he objected to it, was an economic system and an accepted norm in various parts of the world, although it was on its way to being abolished.

My schooldays were not very eventful. I had only a few friends, one of whom was Sayida, the girl I met on my first day at school. Sayida was two years older than me, and gave me a great deal of support. Unlike the other girls in our class, she never referred to my parents' separation. But she left school very early, at the age of 14. She was married to a man twenty-five years her senior and went to live with him in the countryside. Her marriage at such an early age was a shock and a source of sadness for me. I never saw her again.

I finished school and went to university to study Law. Law was not my first choice. I had intended to study either Economics or English Literature, but this would have meant going to Egypt; because I was still young and my mother objected to the idea, this was out of the question. In my Law class I was the only female, and the youngest of 250 students, which put me under a lot of strain. In the second year of my studies I married a lawyer who was a long-standing friend and a colleague of my elder brother. My husband and I agreed that I should finish my studies before we started a family.

After graduating I did my training in my brother's office. I was one of the first women lawyers: at that time, only about eight women had joined the profession before me, and many of those had opted out of the practice either to get married or to take up other legal positions. My experience as a woman lawyer was thus rather unique. Although I chose to work for my brother, my husband had an equally well-established legal practice. When my husband was detained for two years, without charge or trial (under the second military regime which was overthrown in 1985), for being politically active as well as being the Secretary General of the Sudan Bar Association, I spent two years working in his office.

As a woman lawyer I had a great deal of support from my husband, my brother and (needless to say) my father, as well as from my colleagues and the judges. In general, society seemed bemused at having women in the legal profession, rather than objecting to it. In fact, at that time there were distinguished women judges, just as there were women in superior positions in the Civil Service, academia, and so forth. Rather than limiting myself to family cases, I worked in different areas of law. It was interesting to see that only less-educated women preferred women lawyers to handle their family law cases: the better-educated chose the best lawyer, regardless of gender, unless a particularly delicate issue was involved, in which case a woman lawyer would generally be sought.

After a short period of practice as a lawyer I went to Cairo to read for my LLM (Master's) degree. Having finished my studies with distinction, I became only the second woman ever to join the Attorney General's Chambers, where I became a Legal Counsel. I was later chosen to represent the Attorney General's chambers at the Law Commission, which was drafting the new laws, and found myself the only woman on the Commission. As a recognition of my contribution in the Commission, I was awarded a decoration for outstanding achievement in the field of law, by the President of the Republic. However, I had been politically active within the Legal Counsel's union of the Attorney General's chambers, which had earned me the wrath of the regime. Subsequent developments culminated in rigorous polarization of the regime and the popular movement, following an abortive coup attempt in 1971. As a result I was stripped of the decoration.

In 1975 I was granted a scholarship for my Ph.D. at University College London. I travelled over with my older son (Khalid), and my husband and younger son (Amr) followed later. The two boys were then 5 and 3. (Now, twenty years on, Khalid has graduated from the University of Westminster with a degree in Media Studies. Amr has finished law school and is now doing his final training to become a solicitor.)

Having obtained my Ph.D. in Criminal Law, I joined the international department of a leading firm of solicitors in the City. Four years later I decided to start my own practice (Falcon Middle East Consultants). Throughout this period, I have been actively involved in human rights at an executive level in Middle Eastern, Eastern, west European and North American organizations. I have served in the capacity of a founding member of the Union of Arab Jurists,

Baghdad, and in the Law Society of England as an overseas lawyer. I also work for, and promote, women's issues and concerns.

One of my major preoccupations has naturally centred around bringing up my sons in a culture and society different from our own. Although I myself had been exposed to Western culture, languages, etc. at home and through several visits to Europe, living and settling in the West is another thing. The decision has been taken on behalf of my sons. Although we have not decided to immigrate, and still aspire to go back home when things get better, my children have been brought up in Europe. Since their father and I were concerned that they should not lose touch with their culture and language, they had an Arabic-speaking nanny, Mesauda. She was always there for them, particularly when my husband and I had to travel. Mesauda was in her twenties when she joined us; since then, she has become a great friend and a companion to my sons, who love her dearly. She is also a great friend of mine. I taught Mesauda Arabic and encouraged her to learn English. Two years ago she married – she and her husband Nick now have a daughter, whom they have called Safia.

My husband, who had originally started the practice which I am now running, took up journalism and became the editor-in-chief of a London-based Arabic weekly. He now runs his own independent legal practice with a partner. My practice has brought me in touch with, and encouraged my involvement in, women's issues, immigration and sharia law matters. I teach Islamic Law and Human Rights at the University of London, and I write and publish articles on the same topics in professional journals. I have published five volumes of work on similar themes.

It might seem curious, but it remains true to say that the first time I was confronted with open sexism was here in the West. I have not been exposed to discrimination back home. If I had to explain that, I would say that the circles, the profession and the society in which I moved, and my family and class background, must have spared me.

NOTE

1   As long as the man did not own the slave by whom he begat children, the children remained slaves. If the concubine was his own, the offspring (being sons of a freeman) would have therefore been free-born.

9

BREAKING CULTURE'S CHAINS

Safiatu Kassim Singhateh, The Gambia

Maria, 1868–1914

Shuhana, 1896–1945

Balla, 1922–

Sharia, 1948–

MARIA, 1868–1914

Bound by culture

Maria was the adoptive mother of Shuhana. Shuhana was born in Freetown, the capital city of Sierra Leone, in 1896. Her parents were among the few educated and devoted Christians of her tribe, and she was the fifth child of a large family of seven females and one male. They were always referred to by the family name – the family Kankan. The Kankan children were all sent to school and attended regular Sunday classes.

When Shuhana was 3 years old, her parents agreed to have her adopted by her childless maternal aunt – Maria. It was the family tradition to help whoever needed help; besides that, Shuhana was regarded as fortunate to be given to a family where she would be the only child. This was seen as a good omen for an adopted child. Maria was very happy to have Shuhana: she gave her everything a growing child needs, and loved her as if she were her own daughter.

One year after Shuhana was adopted, Maria's husband died. Maria grieved at her loss and mourned her husband's death for a year. During this period she always wore black, and Shuhana would watch her 'mother' weep almost every day. Maria would talk quietly to Shuhana and tell her how much she missed her husband. Shuhana, on the other hand, was much stronger: she learned to listen to her mother and to console her by saying, 'Daddy is in heaven and we shall all join him some day.' Maria always wondered how Shuhana knew about heaven and its relation to death: Shuhana later told her that the priest at Sunday school had taught them about the afterlife.

Soon after Maria came out of mourning, she married a Muslim man from a

122

different tribe. She and Shuhana moved to their new home at the other end of the city. Maria's family were not pleased by her choice of marriage, but because she had already lost a husband they didn't want to add to her grief; they wanted her to be happy.

It was very unusual for families to visit one another when they lived so far apart, and in the case of Maria and her family it was especially difficult because they were now further separated by ethnic isolation. Maria could not visit her own family because her matrimonial family may have felt that she was not happy with them; her sisters could not visit her because they might be seen as interfering in the marriage. And so for several years Maria and Shuhana had no contact with the rest of their family.

Meanwhile Maria had been persuaded to change her religion to follow that of her husband, and Shuhana also converted to Islam. Maria registered Shuhana in the municipal school not far from their house. Unlike the Christian community, however, the Muslims did not send their daughters to school and only educated the boys. Shuhana found herself the only girl in the school. Boys would laugh at her and make fun of her; sometimes they even bullied her. When she fought back they would pull down her wrapper, which formed part of her traditional costume, and expose her underwear.

Eventually, Shuhana could no longer stand this torment and humiliation, so she told Maria about it. Maria made several visits to the school and complained to the headmaster, but all he said was, 'Maria, girls do not go to school, you cannot make a man out of a woman, there's nothing I can do.' Maria realized that it was hopeless to try to persuade the headmaster to protect her daughter from the unscrupulous boys. In order to save her dear daughter from further torment, she took her away from the school and made her do what all the other girls did in the community.

Shuhana was taught how to cook, and how to prepare different types of snacks, including cakes, doughnuts, popcorn and gingerbread, and Maria taught her all the skills she had learned from the sisters at the Christian mission. Maria constructed a stall in front of her house where she and Shuhana sold their produce. They would work in the kitchen and on the stall every day except Fridays, when they had to go to the mosque to pray. In their community, males were placed separately from females, a practice which Maria and Shuhana eventually got used to.

News had reached Shuhana's natural mother that she had been converted to Islam and was no longer going to school. This was an insult to the Kankan family and so Lesina, Shuhana's natural mother, decided to go to her sister to get back her child:

'Oh, it's Lesina, you've come to visit us, how kind of you!' Maria exclaimed as she opened the front door, and embraced her sister.

'Where is my daughter?' Lesina asked, not reciprocating her sister's warmth.

'She is all right. Lesina, is anything wrong?' Maria enquired calmly, holding Lesina by the hand and leading her to a cane sofa.

'Of course there is something wrong,' came the reply. 'I've come to take her home with me, *now.*'

'But you cannot do that, Lesina. The child knows no one else but me,' Maria pleaded.

Shuhana heard the conversation and wondered whether she was the subject of discussion. She stood behind the bedroom door and listened to what developed into a loud argument.

'I am going to the police!' were Lesina's last words, as she banged the door behind her.

For the first time in eight years, Maria and her family met. But it was at the police station. Shuhana was sitting next to Maria and both were facing a row of relatives who had come to support Lesina. The family priest was also present.

The inspector of police listened to the stories from both parties. He found it difficult to make a decision. Maria wept profusely and said that her family was taking advantage of her because of her barrenness. She said she would have nothing to live for if Shuhana were taken away from her. Lesina argued that the child was hers and she wanted her back.

The matter was referred to the magistrate's court. Having considered the evidence before him, the magistrate called Shuhana and asked her whom she wanted to be with. Shuhana ran to Maria, held her closely and said, 'I want to be with my mother Maria, I have never seen the other woman before.' And so it was ordered that, in the best interests of the child, Shuhana should remain with Maria.

SHUHANA, 1896–1945

Restrained in a cultural dilemma

Shuhana was 16 when Maria died of a heart attack, at the age of 46. Maria's second husband Saad had married a second wife and decided that they should both share the house to which Maria had contributed. Shuhana could not go back to her natural mother, because Lesina and the rest of the family had denounced and disowned her after the court proceedings. Besides, she really did not know them; her friends and relatives were within the Muslim community. Saad loved Shuhana very much, but he knew that Shuhana could never accept her stepmother and, worse still, that she could never forgive him for what he had done to her mother. Saad had not expected Maria to be so badly hurt by his taking another wife: it was normal practice within their community and all the other women lived with it.

Saad decided to take Shuhana to his sister Amie, who was also fond of her. Amie was pleased to have Shuhana live with her, but she was not as well off as Maria had been. Shuhana had to make do with far fewer of the privileges she had enjoyed with her mother. She had to help Amie to make ends meet.

Amie had one son who had completed his middle-school education and had left for the neighbouring country of Conakry to seek employment. He had been there for three years. One Saturday morning while Shuhana was busy doing the laundry, Amie watched her from the window. She admired Shuhana, watching as she worked and thinking of all the things she could do so perfectly. Something struck her: this young lady should remain in the family. It was time Shuhana got married.

Amie approached Saad during one of their monthly get-togethers.

'Shuhana has just turned 16 and I want my son Omar to marry her,' Amie said to her brother.

'That is a very good idea, I endorse that,' Saad said approvingly. 'We should send for Omar to come at once.'

'But before that . . . ' Amie said, 'I want to arrange for Shuhana to go through the initiation rites.'

'But Maria never wanted her to go through that, you know; it is not their tradition,' said Saad, with concern.

'Shuhana is now part of our tradition, she is marrying my son . . . and besides, if she doesn't go through the rites of passage she could never be accepted by the community,' Amie pointed out. 'And she will not be happy either!'

Saad had always been concerned for Shuhana and wanted her to have a happier life than her mother Maria. Maria had been regarded among their people as a stranger just because she did not go through this traditional ritual – that was one of the reasons why Saad had been pressured to take a second wife.

So the day was arranged for Shuhana to be initiated. The matter was not discussed with her, nor was her consent required. She did not even know what was involved in this traditional ritual which was enshrouded by deep-rooted mysticism. She spent six weeks in an 'abyss of secrecy'. She came out of it dismayed, fearful and uncertain of what lay ahead of her. Barely two months after her initiation came another blow, when she was told that she was to be married to Omar.

The marriage was contracted and consummated. Omar had returned from Conakry with very little money; the little that he had was spent in furnishing their one-bedroom hut. This time in their lives coincided with a period of severe famine and hardship. One day there was no food in the house – Amie was sick in bed; Shuhana was heavily pregnant with their first child. She was hungry. Omar sat beside her, wondering where he could get some food for his young wife and his sick old mother. There was a mango tree in the compound, which had started fruiting. Omar looked at the tree for a while and said to his wife, 'My dear, I know you are very skilful with your hands. Can you turn these unripe fruits into something palatable?'

Shuhana smiled and nodded. She laughed as she watched Omar climb the mango tree like a monkey and soon there were piles of unripe mango fruits on the ground. Shuhana prepared a good meal for the family and sold the rest: neighbours came to buy mango food every day. Shuhana was the only one who knew how to prepare this special recipe. People had money but no food to buy, so Shuhana was able to trade the mango food and make a lot of money in the process. She felt quite happy and satisfied with her work – not because she had made money, but because she had been able to feed people at a time when they most needed food.

The major events of Shuhana's life affected her personality greatly. Contrary to earlier expectations, she became timid and very reserved. She shied away from people and avoided attending social ceremonies and community events. After several miscarriages and stillbirths, she finally gave birth to a baby daughter. Eighteen months later, Amie died, creating another gap in her life. She persuaded her husband that they should leave the community and migrate to some other place. They decided to go to The Gambia, not too far from Sierra Leone.

Once settled in The Gambia, Omar secured a job with a British firm as a shopkeeper and Shuhana continued to make food for sale. As the business progressed, she imported cola nuts, coconuts, ginger and processed cassava from Sierra Leone to sell to the Gambian community. These items were in great demand.

Meanwhile, their daughter Balla was growing up to be as hard-working as her mother. She learned a number of household and cooking skills from her mother, and her performance at school was also good. Shuhana continued to maintain a very low profile among the Sierra Leonean immigrants. No one knew what was on her mind; she never discussed her problems with anybody. From time to time, she narrated her life history and that of her parents to Balla, who listened attentively and encouraged her mother to tell her more. Balla knew that underneath her mother's conservative and reserved exterior there was a strong personality with considerable willpower. She also realized that this willpower had been overshadowed by some events in her life. Her mother was determined not to allow her own experience to affect the way she was to bring up her own children to face the world.

Balla had several brothers and sisters, some of whom died while they were young. Balla helped her mother raise the four surviving children (two boys and two girls). She completed her primary- and middle-school education and at the age of 19 married a man of her choice.

Through Shuhana's influence, Omar bought a piece of land and built a four-bedroomed house which was quite modern at the time. Balla had moved out of her parents' house to live with her husband, but she often came to visit her mother and to help her with her accounts. Shuhana had saved a lot from her business; Omar also had some savings, but he was too generous and inclined to share his money with friends. He also gave out loans to people who never paid him back. Balla had often cautioned her father about that: 'You have my sisters and brothers to educate, and remember you are planning to return to your home town to build a house.'

But her father would laugh it off and say, 'Child, I am your father and you are my child, too young to see what I can see, hahaha!'

Balla could scarcely remember her father being angry with any member of the family – he was always in the best of tempers. Even when a thief stabbed him when he took him by surprise, her father said to her while he was recovering at the hospital, 'I could have hit him with my big stomach, had he not used the knife on me!', making a joke of the incident (and of his big stomach). Balla and her parents spent some very happy times together. Her mother delivered Balla's first and second sons: she and Omar enjoyed playing with their grandchildren and saw the first child grow up to a toddler.

BALLA, 1922–

'I am the heir of my father'

One day, when Balla was visiting her parents as usual, her father told her that he and his wife had decided to return to Sierra Leone. The primary objective was to build a two-storey house in which they would both retire. He suggested that Balla and her husband should move into her parents' home and give up their own house, which was rented. Balla was very happy for both of them, especially for her mother who had spent all her life working and bearing children. They were going to take the other four children along with them: the eldest was a girl of 14, eight years younger than Balla, and old enough to help her mother. A couple of months later, her parents left for Sierra Leone and she and her husband moved into the family house.

Balla was now expecting her third child. It was a difficult pregnancy and the birth was long overdue. Balla had seen several doctors regarding the pregnancy, but none could give any scientific explanation for the delay; nor could they establish whether the baby was developed enough to be induced. One doctor advised her to consult a traditional medicine man. During this difficult phase of her pregnancy, Balla heard nothing from her parents, even though she had written to tell them about her problem. Her husband Mohamed was very supportive: he asked his sister to come and stay with them so that she could help Balla with the children. Mohamed was a trader, travelling regularly to Senegal and Portugal to sell and buy merchandise.

Meanwhile, Balla's parents had bought a plot of land in Freetown. Shuhana, following tradition, handed all her savings to her husband so that he could arrange for the construction of the house. While they waited for the house to be built, they stayed in Omar's mother's old house which was, by that time, badly dilapidated and far below their usual standard. When Shuhana lamented the inconveniences and the length of time it was taking for her husband to start the building project, Omar would reassure her, 'It is just a temporary arrangement, we shall soon start the construction work, and then it will just be a matter of months.'

As always, Shuhana was very reserved. She didn't have friends or go out, but stayed in the house with her children. She didn't know what was going on outside the four walls in which she confined herself.

One day a friend of her daughter came by and said to Shuhana, 'Aunty, did you hear the news about Uncle Omar?'

'No, what news?', Shuhana asked.

'There are rumours that he plans to marry a widowed woman in the next few days,' the young girl said.

Shuhana did not believe her ears: she could not believe the rumour was true because she trusted her husband so much.

But the rumour was true. History had repeated itself. Omar married the widow and spent all the money that he and Shuhana had saved in the courtship and marriage. He moved out to live with his other wife in a beautiful two-storey house which she inherited from her deceased husband. Shuhana and her children were left in the old hut. Her dreams were broken. There was no money left to build a house.

Shuhana had a nervous breakdown, followed by a heart attack. She died a few months later. Forty days after her death, her husband also died. News of the deaths of her parents reached Balla. She was deeply shocked. She gave a loud scream and, for the first time since the pregnancy had started, eight months ago, she felt the baby move. Balla was rushed to the hospital, and twenty-four hours later a baby girl was born. She was named Sharia.

When Sharia was a year old, Balla went to Freetown to hold a proper funeral ceremony for her parents. By that time, her brothers and sisters had been spread out among relatives who were taking care of them. After the funeral ceremony, Balla's next task was to investigate the situation leading to her parents' death. The story was more or less the same as what she had been told. Next, she made an inventory of her parents' property: there wasn't much. All the money had been spent, and most of her mother's jewellery and best clothes had been taken away. She did not bother to check on her father's personal belongings, as she had already been told that his cousins had inherited all of them. One thing she was particularly determined to do was to acquire the lease document of the plot of land which her father had bought. The land had never been developed. She was informed by a reliable source that her father's eldest cousin had the document in his possession.

Time was running out for Balla – she had to return to The Gambia. She planned her strategy carefully to achieve her remaining goals. She asked that a meeting of the council of elders be called to discuss the disposal of her parents' properties. This idea was welcomed, as a few of the cousins were interested in the plot of land. Balla's brothers and sisters were also considered part of the inherited properties. Even those who were old enough to go to school were being kept at home to do all the housework, while other children of the families with whom they lived (boys *and* girls) attended school regularly.

When the council of elders met with Balla, they showed her the remains of the personal belongings of her parents. Balla took the jewellery for her two sisters and asked that the rest be distributed among the elders and relatives. When she enquired about the lease document, she was told that her father had owed some money and that the property was to be sold to pay the debts. 'I am the heir of my father', Balla proclaimed, 'and I take the unilateral decision on this matter. First, I want to see evidence of my father's debts.' She looked at them steadily: 'I recognize his signature very well.' She waited for a while, then proceeded: 'I'll pay in cash any amount that I find to be my father's debt and which is authentically signed by him.'

It was the first time in the history of the council of elders that a woman – a young woman for that matter – had addressed the elders in such a professional and authoritative manner. Some were amazed, others were dumbfounded and a few felt disappointed and humiliated. 'This is a family affair, I think the woman is right,' the chairman said as he rose to leave the room. The others followed suit, mumbling the words, 'She is right, absolutely right.'

'One more thing,' Balla called after them. They turned round to listen. 'My brothers and sisters appear to be very feeble and sickly. I know you have all contributed to their welfare since our parents died. I want to make my own contribution by taking them with me for a few months for thorough medical check-ups.'

Balla worked hard to raise her own children and her brothers and sisters. It wasn't an easy task. Her husband's business was not doing well, and they could not afford to meet all the basic needs of the family. Balla was obliged to start a trade of some kind. She tried the fast-food business, as her mother had done, but it was too hard for her: the heat of the open fire, the trip to the market every day to buy fresh foodstuffs, carrying firewood – it all became too much. She hadn't been brought up to such a hard way of life.

So Balla decided to sell fresh vegetables. This was much easier, as she only had to get the vegetables from the market three times a week. She did not make as much profit as her mother had done before her, but it was just enough to subsidize what her husband brought home. She spent her free time helping the children with their schoolwork. She was particularly fond of telling them stories and playing puzzle games with them: she hoped this would help cheer up the children, especially her orphaned brothers and sisters.

Up to this point, Balla had had quite a good relationship with her husband, Mohamed. They supported each other in many ways, which made life easier in terms of family responsibility and decision-making between members of the family. Balla maintained a high profile in her community and an equally strong authority within the home. These qualities were not common among women of her age group. Even though she had spells of emotional stress while raising so many children, she was strong and coped well with her problems. Unlike her mother, she had not been exposed to some of the deep-rooted traditions, the emotional and psychological effects of which had made her mother unable to deal with stress: her mother had told Balla that some of these practices have no meaning in one's life. Balla's strength was further enhanced by the fact that she had inherited some property from her parents, one of them being the house in which they lived. This degree of economic independence enabled her to resist cultural pressures and some of the unacceptable suggestions that her husband made from time to time. So, when Mohamed was pressured into taking a second wife because Balla had decided not to have any more children, Balla asked him to stay away from her and her house. This sanction was hard for both of them.

Balla was not interested in another marriage. Her main concern was to ensure that her siblings received a good education and got the support they needed to face the problems of the world.

Balla was reconciled with her husband after his second marriage was dissolved, when their youngest child was 8 years old. Balla had managed to improve her business, and her economic situation was far better than her husband's: this gave her complete control over herself, her home and everything that was associated with her life. She had a good sense of judgement and she passed this on to her children by guiding them into taking proper decisions.

SHARIA, 1948–

Willpower to change

One of Balla's daughters, Sharia, grew up to be really outstanding. She did well at school, was assertive and outspoken. Her parents were very proud of her. The aunt after whom she was named, however, considered these 'so-called qualities' as obstacles to her future. Aunt Sharia was a traditionalist and believed that young Sharia should be 'helped' by crushing those unusual qualities out of her. Sharia's parents never took Aunt Sharia's comments about her namesake seriously, and would laugh, disregarding her opinion totally.

One of the family traditions that Balla and her husband valued was to send each child to her/his namesake, wherever they may be, to spend one summer holiday with them. This holiday was a privilege for each child, granted only if she/he completed a high-school education. Such a visit would therefore happen when the child was between 17 and 19 years of age. Eventually it was young Sharia's turn to go on holiday to her aunt Sharia. This time spent together had great significance for both parties: it was a period when the two became close to each other, when the older person could assess the personality traits of the young namesake and attempt to pass on certain characteristics, while the younger person was expected to be observant and to demonstrate her/his best qualities. What might be considered 'best qualities' was to be determined by the older person.

After spending a week with her aunt, Sharia was invited to meet some other members of the family, whom she had never met before. She was told the visit was a special occasion and she looked forward to it. Sharia noticed that throughout the ride to the meeting place, her aunt and the two other women who accompanied them spoke in a language she did not understand – the

Indigenous language spoken by her paternal tribe. As was her nature, she was curious to know what they were talking about. She wanted to be part of the conversation, but she knew that whatever was being discussed was meant to be secret to her, which was why they did not speak in the language that was common to them all.

The car stopped at a house which was well secured, with a tall, tightly woven bamboo fence. Sharia could not help commenting on how secluded the house was, situated among some giant-sized baobab trees. The nearest house she remembered seeing was almost a kilometre away.

As they entered the gate, they were met by seven middle-aged women, many of them tall and stout. 'This is Sharia,' her aunt introduced her, tapping her on the shoulder. Sharia was about to curtsy as a sign of respect for the women, when she was suddenly picked up by fourteen strong hands and carried into a big hut at the back of the yard, where other women were waiting.

What happened in the hut was the greatest nightmare Sharia had ever known in the nineteen years of her life. She couldn't imagine that anything worse could ever happen again, either to her or to anyone else in the world. She was later told that she had had to go through this initiation ceremony in order to redirect her ego. The decision to put her through this rite of passage was taken unilaterally by her aunt.

The bitterness and anger aroused in Sharia led her, over the following years, to consider how to address some of those aspects of cultures that her people believed to be useful and protective. Members of her community could not understand that some of these practices have adverse effects on the lives of the victims, because no one was willing to discuss the issues involved: it was forbidden to do so. As Sharia had to tell herself over and over again, when she thought of her experience, there was no use crying over spilt milk. The damage had already been done. Her task was to rebuild whatever had been destroyed in her and to work towards the dismantling of certain traditions.

Sharia worked her way through college and university and worked in both the public and private sectors. Throughout her work experience and in her interactions with people at the national and international levels, she has continually raised these cultural issues and encouraged public discussions and debates on them. She has tried, in various ways, to influence policies that have some bearing on such cultures. She has left no stone unturned, mobilizing educated as well as non-educated women and men to take up the campaign against cultural practices that have no meaning in their community. At the same time she has launched promotional campaigns on those cultures that are enhancing to humankind and society as a whole.

Sharia is married and has children. She has used the experiences of her mother, her grandmother and her great-grandmother to build a different world for her own children and the children of her society.

REFLECTIONS

The lives of the women in this four-generation story represent a set of experiences which, if critically analysed, may create a process through which women of today could understand the dynamics of their culture and those factors that inhibit and/or enhance their lives.

Stimuli

Chufelle: loving and caring

The concept of *chufelle*, or loving and caring, can be conceived and expressed in relatively different ways. What is considered an act of loving and caring by one person may be interpreted differently by another. These differences in expression and conceptualization are determined by a number of factors, such as level of knowledge, experience, attitude, ignorance and anxiety. Thus the implications of 'loving and caring' may have significant variations. What is important is how the individual is able to deal with whatever consequences the act of *chufelle* may have on her.

Maria The story does not tell us much about Maria's childhood life. We know that she was born into a well-off family, and a large one. Her parents sent her to school just like her brothers and sisters. Her education at a Christian missionary school indicated her parents' association with, and belief in, the Christian religion: educating one's children and bringing them up within a religious background was (and still is) a way of loving and caring. It is a way of investing in human capacity, in preparing one for the present and the future. This was just as true in Maria's time as it is in our own. Another expression of love and care for Maria was demonstrated by her sister, who gave one of her own daughters to relieve Maria of the pain of barrenness. This invaluable gift may have enhanced Maria's love for herself, as well as for her adopted daughter and for society as a whole; it may have contributed to building her strength and confidence, and her determination to empower herself economically. The latter quality was manifested in the use of her acquired skills for commercial purposes, as well as in her desire to transfer such skills to her daughter Shuhana after Maria's unsuccessful attempts to educate Shuhana through the formal school system in a community which was against female education.

Shuhana Similar expressions of love and caring were passed on to Shuhana by her adoptive mother. We may recall that Shuhana was converted to Islam after her mother married a Muslim and moved to a predominantly Muslim community: this conversion to another religion was a way of belonging to and being accepted by the community whose norms and values governed their day-to-day lives. We have also seen a different expression of love and caring shown to

Shuhana after the death of her mother. Her stepfather's sister initiated her through the rites of passage so that she may belong to the community and, most importantly, she arranged that Shuhana be married to her son. What have all these meant for Shuhana? The story told us that soon after the initiation rites Shuhana became withdrawn, fearful and dismayed at what had happened.

But how did Shuhana judge her aunt and stepfather who put her through this, and how did she assess herself? As readers of the story, we cannot know the answers to these questions. What we do know, however, is that Shuhana loved her mother-in-law, her husband and the community, as was clearly seen during the period of famine. She was able to sustain the strength and determination which had been implanted in her by her mother, and to use these to survive. It was through her initiative and hard work that she and her husband were able to migrate to another country, where they developed their lives. Shuhana's interpretation of the kind of love and caring given to her during her childhood and adolescent life – not only by her immediate family but by the community at large – was determined by her conceptualization and understanding of situations as they arose, and by how she dealt with these situations in the living of her life.

Balla and Sharia   These two characters grew up in somewhat similar circumstances. They were both loved and cared for in ways that probably differed from their grandparents and great-grandparents – different in the sense that modernization and social mobility were having an influence on norms of parental love and care, and community concern for individuals and members of the family and society. Loving and caring were based on needs and wants in a more flexible and reciprocal manner. This period saw much play on the dynamism of knowledge and experience, which in turn led to a wider variance of attitude. In Sharia's time we see certain forms of expressed love being challenged openly rather than dealt with silently, as in the case of Shuhana. Balla asserted her rights to the property of her deceased parents by challenging a group of elderly councillors and taking custody of her younger siblings. To the elders this was an act of insubordination and shame, showing disrespect for elders; but to Balla it was an act of loving and caring for her late parents and for her brothers and sisters.

Kolare: respect, appreciation, self-esteem

Kolare literally means 'respect for and appreciation of others and self-esteem for oneself'. *Kolare* can inspire one to attaining achievements and successes in the most assiduous and scrupulous ways; but it can also lead to a state of despair by being too tolerant and compromising. The four characters in this story demonstrate *kolare* in its positive expression.

This is shown, for example, by Maria, when her sister Lesina and her whole family turned against her, taking her to the police station because of the conversion of Shuhana and her withdrawal from school. Maria, according to the story, did not express any form of indignation or anger at this – she understood and appreciated the reasons for her sister's reaction, even though being summoned to the police station (especially by a family member) is a serious stigma for anyone to endure. Most importantly, Maria respected her sister for the great gesture that she had made by offering her child to her in the first place. Lesina and her family, in their turn, could have appealed against the decision made by the magistrate; however, doing so would not only have increased bad feelings among the members of the family but would also have worked against the interest of the young child. *Kolare* was therefore applied by both parties in settling the matter.

In the case of Shuhana and Sharia, the rites of passage imposed upon them, although initially creating feelings of disgust, bitterness and dismay, were translated to reflect a reciprocal feeling and desire to protect and to harmonize relations rather than to harm and hurt. This explains why both these characters maintain good relations with their 'assailants'. In Sharia's case, this particular experience gave her the opportunity to work towards an understanding within society of the implications of this practice, so that *kolare* could manifest itself in the process.

Honesty and trust

All the women in this story showed honesty and trustworthiness in their dealings with their partners, at their work and within the community in which they lived. They seemed to see this as an important factor for maintaining cohesion within the family unit and the society. Shuhana, for example, worked very hard to improve the family's economic position with the hope of establishing a better lifestyle for them all. Her trust in her husband was evident when she entrusted him with all her earnings and allowed him the leading role in planning their lives: she realized too late how naïve she had been. Honesty and trustworthiness have their merits, but can only work well when they are reciprocated.

Constraints

The constraints that cut across the life histories of these four women can be categorized as follows.

Conservatism

The time and the society in which the women grew up were characterized by a general desire to maintain the status quo, to retain whatever has been put in place by history. This of course meant the perpetuation of delusions and resistance to change, which imply the non-acquisition of knowledge and experience and therefore no change in attitude. For each of the four, however, the degree of conservatism varied from one community to another within a given time frame, as well as from one generation to another, and diminished gradually over time.

This confirms the dynamism of cultures, values and societal norms.

Complacency

This characteristic was seen in Shuhana after she and her husband returned to Sierra Leone. It seemed as if *kolare* for her partner and herself was having a negative effect on her and her children. She was too tolerant and too compromising, to the detriment of her own self-esteem.

Fear of stigma

By not doing what others do, one becomes the 'odd person out'. Women may fear stigma even if they are convinced that what they are doing is the right thing. This fear seems to have played a part in the minds of several of the characters in the story.

Other constraining factors include ignorance, lack of education, poverty, and familial and societal pressure.

Choices

In terms of choices faced by women of today and tomorrow, it may be said that the late nineteenth and early twentieth centuries had some disadvantages: opportunities for women and girls, in particular, were entirely dependent on the roles that society defined for them, the way they were socialized in terms of their female gender (as opposed to the male gender), and the relationship of each to the other. The options for females were severely restricted, as the level of awareness in society which could have enhanced the chances for both human and technological development was equally limited. In spite of these obstacles, this four-generation story has shown the women concerned making enormous efforts and undertaking innovative actions to make the best out of their situations, and especially to cope in times of crisis. On the other hand, we have also seen that individual efforts can only be meaningful if society understands, appreciates and supports such efforts.

By contrast, our situation today is graced by the continual advancement of technological development, communication, dissemination of information, and social mobility. All these have facilitated increases in knowledge and awareness, changes in attitude and therefore in ideology. Taking this into consideration, one can highlight some of the choices faced by women today in shaping their realities.

Taking advantage of emerging situations

In my society today the myth associated with the education of girls is fast eroding, even within the rural communities. More women are found in the teaching profession, especially in early child education and middle schools. This gives women great advantages in sharing the realities not only of their own lives but the lives of the next generation. Within the educational institution, in the minds of the children the teacher's ideas often supersede those of the parents. Children believe in their teachers and look upon them as mentors. Institutions of learning have always been responsible for creating and sustaining ideologies and values: it is important, therefore, at this point in our process of development, to consider how women could influence change within institutions of learning.

Within social service programmes, too, women form the majority of workers, including nurses, midwives, community development and social workers. Their interaction with women at grassroots level provides great potential for empowering women with knowledge and information that would enable them to establish new and positive values and attitudes that would give meaning to their lives.

The challenge facing us, as women, is to understand the origins of our subordination and our roles in society in relation to our male counterparts, and to develop a concept which could form the basis of an ideology that is responsive and friendly to the female gender and which provides for equal opportunities for both male and female. Such an ideology should direct the socialization process within the home, the family and the society, and should influence the contents of our curricula in all institutions of learning and training, at religious centres, and at all levels.

Making the best of what we have

While we strive for change, the reality is that we will, for quite some time, have to live with our current condition and status as women. Our challenge in this respect is to make the best of what we have. For instance, all women, be they in rural or urban areas, upper-class or middle-class, educated or illiterate, share many similar problems based on *cultural values*. Some of these cultural values relate to other factors within the environment, and it is the type and nature of these other factors that is responsible for the wide variety of situations and conditions in which women find themselves. Closely related to this is the level of *development* or, to put it more crudely, the level of *civilization*. It could therefore be suggested that there are indeed similarities among females in the ways of shaping their lives. What is important is the knowledge and understanding of the 'dynamics of the game', and the willingness and the effort to share such knowledge and experience through interchange and interactive exercises and a continual analysis of the processes.

10

DEFENDING THE *SAMAY*

Christina Gualinga, Ecuador

Maria Magdalena Santi Tanchima, 1881(?)–1966

Elena Ruzmila Gualinga Santi, 1927–

Christina Gualinga, 1962–

GRANDMOTHER

My grandmother is among our ancestors

The state in which I was born is officially known by the name 'Ecuador', a country with ten million inhabitants, of which Indigenous peoples make up the great majority. It is situated in the continent which the Europeans named Latin America. But the nation my people belong to carries a different name. It is one of the nine Indigenous nations of the Equatorial Amazon, each of which has its own customs, religious beliefs and ways of life. These nations consist of more than four hundred communities. The language of my community is Quechua, which is spoken by the majority of Indigenous people in Ecuador.

My grandmother was born at a time when the people of my nation did not register the comings and goings of human beings in terms of dates, according to the Western system. So I do not know the date of her birth or death. She passed into the world of our ancestors when I was 3 or 4 years old. I vaguely remember her face on her deathbed. My mother told us very little about our grandmother as a person. In our community one does not talk about oneself or, in general, about details of the lives of individuals, but we are told a lot about the history of our nation and the ways of life of our ancestors. Continuity comes from that history rather than from the history of the nuclear family. But it is not true – as some anthropologists like to suggest – that the group is so important that the individual does not count. In our community private life is as important as community life. The two are complementary. But the story of the lives of individuals is usually told through stories of the life of the group.

Thus I was told that long ago, in the age of our ancestors, life was much more closely in harmony with nature: there was no need to wear clothes, because the

138

forest covered our nakedness. Later, in the age of our forefathers, we experienced the first encounter with people from overseas. At first my people did not realize that they had come in search of gold and other precious things that would make them rich in their own part of the world. My forefathers received them graciously according to our customs. In the name of what they called 'civilization', the outsiders started to claim our ancestral lands. My people did not understand what they meant when they said, 'And now this land is our property', because in the conception of my people one can never *own* the earth. In our language we do not say this land is 'mine', but this land is 'for me', which means that I can till it, but not possess it.

The Spanish conquerors were followed by the Catholic priests: Jesuits, Salesians, Franciscans . . . Different religious orders, some with white clothes and some with black, began to enter into competition to 'save the souls of the Indigenous peoples': my father called this the battle between the 'yurac fathers' (white fathers) and the 'yana fathers' (black fathers). They invaded the Indigenous territory with the purpose of 'evangelizing' and 'civilizing' the Indigenous people.

This brought much suffering to men and women alike. The Indigenous women used to wear short skirts, but when the Catholic priests settled in our villages the women had to quit their traditional clothes, because the priests perceived them as a provocation to men – although our men had never taken offence at the women's ways of dressing. But the Catholic priests forced our forefathers and foremothers to accept many other things too – their religion, their language, their education – all of which were completely alien to what constituted our vision of life and dignity.

They made it an obligation for the Indigenous population to attend Mass every Sunday. If someone did not turn up, she/he was punished in public. The priest would use his long sharp fingernails to pinch the earlobes of the person concerned until they bled. They also changed the system of marriage. Before the Catholic priests came, marriages were arranged by parents, but the boys and girls were consulted. The Catholic priests, who wanted to replace the authority of the parents, changed this: the men of the village had first to go to the priest to tell him about their wish to marry. The next Sunday, during Mass, the priest would ask, 'Who wants to marry?' and the man would answer, 'Me!'. The priest would then call the man to the altar and ask him, 'Which one do you choose?'. The man would point at the young woman he had in mind, although he often did not even know her. The woman then had to come forward and the marriage took place immediately. The woman had no choice: if she did not want to be married and tried to slip quickly out of the church, she was brought back and forced to accept the marriage; the priest considered the desire of the man sufficient to perform the marriage ritual. This is how my grandmother was married.

She had quite a number of children, but my mother did not tell me how many. Several died very young. One of them is still alive, however: he is a shaman, a traditional healer. I remember visiting him just once. He lives far from my village – three days by canoe on the river Rio Probomaza. I have been told that it is very difficult to become a shaman. The apprentice is taught by an elderly shaman for many years; he has to follow a very strict and sober diet – no salt, only some fish, manioca and plantains – and drink a liquid called *jayahuasca*, which is extracted from a plant. By the time he comes out of his initiation, the apprentice shaman is very thin.

MOTHER

My mother is among my father's people

My mother was so frightened about the stories of the forced marriages that she took the risk of deciding for herself to marry my father. Her parents did not agree, but she left with my father for his village, as was the custom, and has remained there ever since. She saw her parents and her brother very seldom, as did we, her children.

At the time of her youth, the Catholic priests used to insist that the boys went to school, but not the girls, so my mother does not know how to read and write; nor does she speak Spanish. But the elderly women of her village had shown her everything that a woman needed to know in order to care for a family and for the ancestral land. She learned how to sow the *yuca* (yucca), which is the basic food of the Indigenous people of the forest. To sow such a plant is a real cere-mony, a process which respects the vital cosmic energy according to our old traditional rites. The women of my people knew long ago about what the indus-trialized countries of the West nowadays call 'biodynamic agricultural production' and 'homeopathic medicine', because for ages that was their way of caring for the people and the land.

My mother knew a lot about all sorts of forms of life in the forest and also about those which have disappeared: the water creatures, the *yacurunas* from the forest, the *sacharunas* of fertility. She knew about planting, sowing and reaping. She also knew about the animals which were the equals of the *Supay*, which are the gods, and about the deities which preserve the life of the forest, of the rivers, of the lagoons . . .

I witnessed my mother's knowledge of the forces of plants many times: for instance, when my father woke up at night from a dream. My parents and the five children all slept in the same room; until the age of 6, the children slept in the parents' bed, after that they moved to the children's bed. Dreams are consid-ered to be of great importance in my community. Dreams are like a newspaper, telling of events in the life of the community, and also in the larger world. It is said that in dreams you travel, which means that through dreams you know what happens elsewhere. Dreams were always communicated and interpreted. When my father had a dream in the night, he used to wake up my mother. She would get up and make a fire to boil water: this woke us all up, and we then sat at the fire listening to my father telling his dream. My mother made a special tea for him from the *huayusa* plant which she had collected in the forest. There were two ways of preparing the plant: one for bad dreams and one for good dreams. After a bad dream, my mother prepared the tea in such a way that it would make my father vomit and so clean his stomach. After a good dream, he did not vomit. She also knew about other teas. When my father went out hunting, for instance, and could not eat all day, she made a tea that reduced his appetite.

When my mother had a dream herself, she would not get up and make tea. She simply stayed in bed and told her dream. It was not the same ritual. It seemed that my father was treated as the most important person in the house, although my mother did the hard work. She took care of the children, the household and the food production. She never questioned this division of tasks. There was no discussion about women's and men's tasks: it was simply like that.

The relationship between husband and wife is a very delicate and complicated topic. Indigenous men are naturally very jealous – perhaps for reasons of insecurity or fear, or perhaps simply because they see women as very delicate, like fine crystal vases. What I do know about the relationship between my parents is that my father was a courageous man, but also very jealous. He would never go out to meet friends without my mother.

Relationships with friends also play an important role in the life of a couple; a friend is a person with whom we can share our experiences and our ideas and also discover other experiences; often in meetings between friends, the experiences of husband and wife are dealt with. My father was 'macho' in a positive sense. He loved my mother so much that he did not want to take any risk of other men liking her too. This makes me believe that my mother was really a part of his heart. My father was not exceptional in the community: many men felt this way about their wives, with the result that extramarital love affairs were very rare.

During meetings of the community the women were all present, although they spoke less in public than the men. However, they spoke with their husbands before the meeting, so that the opinions of the women came through indirectly. My mother and the other women did not seem to mind this situation; they felt that they were, after all, the most important people, since the men could not speak at the meeting without having first consulted with their wives.

This situation has now changed, like everywhere else. Women as well as men have been trained for leadership, to stand against the oppressors. The Indigenous struggle is also the women's struggle.

CHRISTINA, 1962–

I am among my husband's people

I was born in 1962 in my parents' village in the Equatorial Amazon. If you were to ask my parents how many children they have, they would say six, because in my community you count the number of pregnancies. One of my sisters died when she was 2 years old, but I still have one sister and three brothers. I am the youngest child.

I remember my early youth as a wonderful time, enjoying the freedom of playing in the village. My brothers had gone to school, my elder sister not. But the ideas of the Catholic priests about girls going to school changed during the 1960s, so, when I was 6 years old, the Catholic priest came to tell my parents that I should go to school. My father said 'No', because he was afraid I would have contacts with people outside our community. My mother also said 'No', because she knew I did not want to go. The priest needed four years to convince my parents and me. Finally I went to the village primary school when I was 10 years old.

A year later, the priest came again to tell my parents that I should go to secondary school in the city of Puyo. I could stay with the Catholic sisters there and they would take care that nothing would happen to me. At that time I was still sleeping in the same bed as my parents: I was very frightened of leaving home. But I had to go. To get to Puyo from my village meant travelling by canoe for three days, so I could go home only once a year.

I was put up in a sort of convent of German Catholic sisters. It was surrounded by a wall and you could leave through only one door. It was a prison; simply horrible. I felt as if I did not exist any more. There were four other girls there as boarders: we were treated by the sisters as objects, not as persons, and certainly not as personalities. We were the objects of their rules: as such, we had to respect the very strict regulations of the boarding house, go to Mass every Sunday, go to religious classes all of Saturday. Joy seemed to be a sin. We did not have the right to leave the walled garden except to go to school. We did not have the right to participate in the community life of the sisters. We did not have the right to prepare food in the kitchen. Our only right was to study.

Slowly I became aware that in the outside world something was going on with my people. I heard talk of the oppression of the Indigenous people and the creation of Indigenous movements. I realized that what the German Catholic sisters were doing to me now had been done to my people for ages. I started to think that I was really wasting my time with the sisters.

One day the other four boarders and I were speaking in Quechua amongst ourselves, when one of the sisters told us that it was strictly forbidden to talk Quechua. When I asked why, she answered, 'Because it is better for your Spanish.' I asked whether this was the only reason. She said, 'What's more, we don't understand.' I think this was a slip of the tongue, but I became furious and shouted, 'Then you should not talk German, because we don't understand your language either!'

This episode started a war between the sisters and me. I decided to leave, but of course I had to do that secretly. I took my bag with all my books, because I wanted to continue to study. I climbed over the wall and walked through the city to find the house where my sister was working as a maid. I was frightened about my newly acquired freedom, but proud too.

My sister was working at the house of a lady who was a teacher at the same secondary school. At first my sister was also frightened, but then she decided to help me. She decided that with the little money she earned we could rent a very small place together. My brothers came to help out, too.

While staying with the German sisters, I had contracted tuberculosis. After a year I was so weak that I had to go back to my village. When I was cured I returned to Puyo, where I finished secondary school. After that I started to work in the organization of the Indigenous movement. I was first secretary of the Organisation de Populacion Indigena de Pastaza, which had all sorts of projects: education, agriculture, territorial rights for Indigenous people, and so on. After two years I became the bookkeeper for the international projects: and that is how I met a Belgian development worker who became my husband. When his contract was finished we returned to Belgium, where we are now living with our two children.

My parents have simply accepted my choice. They never questioned my private life. My eldest brother is married to a girl from the village and the other two brothers to women from other communities.

Coming to Europe has been a very profound change for me. I am glad that I have had to learn to appreciate different cultures, not to shut myself up on my own and not to reject things. I also feel that, since I am here, I should use the opportunity to raise awareness among Europeans of their own culture, of the damage that their kind of 'development' is doing. When I talk to groups here, I think it makes more of an impression than if a Belgian were saying the same things. I feel quite content here in myself, but less happy with my contacts. People here are more reserved, colder. It is a different way of life and a different way of communicating.

Since living here, I have also come to understand much more clearly how the oppression of my people works: when you are in the village you cannot see how the outside world functions. Now I realize that the oppressors are the State and the oil companies. What they mean by 'development' to us means destruction: destruction of our forests and the world environment; destruction of our cultures and those of other people who respect nature; destruction of our beliefs, religion and customs. In one word, destruction of our dignity.

In fact, my language does not have a word for 'dignity' in the Spanish sense. The appropriate word would be *samay*, which means a number of things at the same time: to breathe in a spiritual sense, to live in harmony with others, a pure life in relation with nature, but also in relation with the past, the present and the future. To keep the *samay* is at the core of what Quechua people are seeking. It provides us with tranquillity, security, sanity, strength and calmness. We have to oppose everything that destroys the *samay* – bulldozers, armies and pollution all destroy our spiritual and physical 'breath', which are one.

I learned from my parents that the Earth is our great Mother, our *pachamama*. The Earth is the great wholeness. When we speak about our Earth, we say '*ñucanchi huiñay causana allpa*', which means 'the Earth where we will always live' – before, during and after our human life. And what we say in our language, Quechua, finds an echo among other peoples and other nations which share this vision. I was taught at home that the Earth is not an object which we can possess and reproduce, it is not some merchandise which we can appropriate, but a material support in which we should freely delight. We have an important and special relationship with Mother Earth, which is essential, vital for our life and for our beliefs, traditions and culture. '*Rucuguna huiñay causana pachamama, ñucanchi huiñay causana pachamama*' – the Earth of our forefathers, our Earth, is not only a part of the forest which the laws of the State of Ecuador call *baldias*, referring to the territorial surface. It is far more than that. As our forefathers have taught us, it is also the *Ucupacha*, where the *Supay*, the gods, have their homes and where their *Yachai*, the protective forces, take care that the life of plants, animals and human beings will grow.

It is in the *Jahuapacha* that the cosmic vital energy and other forces enter: the rain, the clouds, the winds, the flashes of lightning and the thunder. Here the lives of nature and of men meet. When we speak about our territory, our Earth, the *pachamama*, we refer to the whole universe. This is why, for our people, to defend the inheritance of our forefathers means to avoid the destruction of the forces that are an integral part of our territories.

Western society presents itself as the Saviour, offering a kind of paradise. It fixes its eyes on the Third World under the pretext of 'developing' us; but all the while it is evacuating the waste produced by Western industry (toxic, radioactive waste) to the so-called 'underdeveloped' countries. This behaviour is not worthy of human beings. At the same time, it is creating centres of prostitution in the forests to cater for the workers from the oil companies. In our peasant communities, there are no hospitals for treating illnesses that come from outside our reality, such as AIDS, and other sexually transmitted diseases (although we are not – yet – in such an extreme situation as some parts of Africa).

'Modernization' has also brought about a shift in the long-established roles of men and women, and this too has created destabilization. In our communities the women are taking on more and more political and social functions and, because of this, are facing problems in their marriages – troubles that can often damage the whole family. This is also happening partly because men do not yet realize that women are able to divide their time and space in order to organize their work outside the home. I support those women who have undertaken a concrete task in society and therefore have to combine their work and family life.

In the Indigenous society, there are special tasks which can only be done by the husband, such as hunting and the preparation of traditional medicines. This division of tasks is based on complementarity with nature. It comes from well-established traditions, and as women we do not feel inferior because of this. Nowadays, in our society, some people think that women are in search of work outside the house in order to avoid their responsibilities at home. I do not believe this to be true: it is just that we want to contribute with our knowledge, we want to share our experiences, meet with others who share the same points of view.

144

The contributions of men and women are complementary to one another both inside and outside the family.

The woman who also functions outside the home is more responsible, more concerned and committed, and does her job with more enthusiasm. I believe that we can only change by unifying our human forces, valuing the efforts and the work of both men and women, making no distinction between the sexes. Since our social system is not used to the idea of such values, we find ourselves living in a society that creates uncertainty and does not satisfy all personal needs. I am particularly concerned about the women who have made prostitution their work, and are, because of this, excluded from normal social life. Unfortunately, however, no other solution has yet been found to tackle the emotional problems of men, while at the same time meeting the economic needs of women.

I think that, while political discourse must be analysed deeply, we have to switch to action. People must express themselves, must talk openly about truth and justice. We must call on the conscience of the world population to demand the purification of what is already polluted, to stop any further pollution and destruction. We cannot take the responsibility for letting future generations suffer from the consequences of this 'modern' society which has created so-called 'development'.

11

TRAVELLING IN A GREEN STONE

Eliane Pontiguara, Brazil

Maria de Lourdes, 1916–

Elza, 1928–

Eliane, 1949–

THREE LIVES INTERLINKED

An Indigenous ghetto

My grandmother, Maria de Lourdes, was born in the Indigenous Nation of Pontiguara ('he who eats shrimps'). Her father disappeared during the violent process of colonization. In 1927, at the age of 11, she became pregnant as a result of chauvinist patriarchal violence, and in 1928 my mother Elza was born. In the 1920s, in a traditional Indigenous village, it was an enormous shame if a young girl became pregnant by a foreign man. My grandmother's family (four sisters and one brother) and my mother emigrated to Recife and later to Rio de Janeiro. An Indian called Mr Marujo (Sailor), a very old and ill man, almost blind, whom I met in 1979, still remembered the sisters Maria de Lourdes de Souza, Maria Isabel and Maria Soledade, daughters of the old Indian Chico Solon and Maria da Luz Solon de Souza. Chico Solon, my great-grandfather, died forgotten. We never heard of him again. Only a few old men will talk about him if they're asked. At that time the foreign colonization was very strong. The old men say that in those days those who were against the oppressing system were put in a bag, shot and then thrown into the sea with a rock tied to their feet.

My grandmother was a combative woman and fully conscious of her reality as an Indigenous woman. In spite of suffering discrimination wherever she went, for being a woman, northern and Indigenous, she conducted her life with dignity. When she arrived in Rio de Janeiro (already with three children), she found a house in the immigrants' zone near to the prostitutes' part of town. Abandoned by her husband (an insensitive and violent person), she had to suffer the worst humiliations, both moral and physical. With a lot of hard work, she raised her children (three girls and one boy). Unfortunately, while still very

148

young, my mother married a hawker, a very violent man who had relationships with several women. I was born into this marriage, and lived under these conditions until the age of 6, when he was run over by a moving tram.

After that my mother, my brother and I went to live with my grandmother, who was still living in the immigrants' zone, by the railway station in the centre of Rio de Janeiro. My grandmother and my mother were extremely worried about my future, so they locked me inside the house. I had no access to things of childhood or adolescence; I never played with other children, except my older brother, and I could never share my dreams or fantasies with other teenage girls.

My grandmother created an Indigenous ghetto inside the house and we lived in a cultural cell. I was not conscious of the process at the time, but years later I realized what those women had done to me. I was ill until the age of 20, suffering chronic anaemia and tuberculosis because of the lack of sun. My world was limited to a small house with one bedroom and one living room. At night I was always scared because I slept in a room over the basement; I could hear the rats in the basement and under the bed.

I used to watch as unexplained tears rolled down my grandmother's face. Many times I saw her drinking pure *cachaça* (a very strong liquor usually made of sugar cane), deeply unhappy. With me she was always very kind and tender, but demanding; she wanted me to study. She didn't like me to talk to anybody and demanded that I shouldn't look at anybody either. She would take me to school and forbid me to talk to the other children. I had problems learning, and I used to cry a lot.

In our house there was not much food, and my mother and grandmother used to give me all the milk and meat instead of taking any for themselves. I couldn't understand why I was the only one who had milk and why I always got the best food. It used to give me stomach-ache. One day my grandmother gave me a green transparent stone. That green stone was the link between generations in my Indigenous family; with it, I received all the culture, all the tradition and all the spirituality that these women carried all their lives. I used to travel in that green stone. That was the positive side of my maternal house-prison – it was in fact my Indigenous village, in exile from its own land.

I spent my childhood and adolescence witnessing the silent tears of my grandmother. She never talked about the past. I discovered her past and I glorify and honour her. I think that I am the result of an Indigenous culture and education that were kept, like an island, inside an old flat, urban but full of wisdom. It was beautiful – the wisdom of that woman who knew how to live a life ruled by her own experiences, her own work and the hope of building a different world for her children and grandchildren. Hope and dignity motivated her towards the future. My life was dominated by the presence of my mother, aunt and grandmother, all strong, decisive, combative and lonely women. My aunt later married, but until then she shared the loneliness of her mother and sister.

The female lion preserves its young, and Maria de Lourdes and Elza preserved me as if I were a diamond. That's why they locked me in a flat and did not allow me to communicate, even in school. My mother, who suffered more directly the impact of immigration, of the humiliations – of poverty, of hard physical work, of diseases, racism, unhappiness, the lack of childhood and friends – was always very hard and bitter, showing unhappiness but also optimism and a strong will. With those women, I learned how to fight and win. I never learned the word 'defeat'.

My mother has suffered a lot during her life. She has twice been widowed. In spite of her strength and her love for the Indigenous cause, I see that she is very fragile and unhappy. She was deeply influenced by her mother, and she lived the lives of those who surrounded her – her mother and her children. Indirectly, my grandmother motivated me to fight for the Indigenous cause. At home our eating and cultural habits were Indigenous, eating roots and breadfruit. In our relations with society we were Indigenous: I saw the discrimination towards my family and myself.

My grandmother couldn't read or write, and my mother and uncles received a very basic education. Through the effort and the sacrifice of my family, I managed to finish a higher-level course in Education and Literature. But my personal life has also been marked by suffering. My husband asked me to choose between him and my Indigenous struggle: I stayed with the Indigenous struggle and was left with three small children to raise. (My husband died recently: at the end of his life, he came to recognize my struggle, but his understanding came too late.) I lost everything and went to live in a shanty town (*favella*). With my wages as a teacher, I managed to get back on my feet after about one and a half years, and we moved to the centre of the city. I had the support of three people – Herbert de Souza (Betinho) from IBASE, Pereirinha (a trade union worker) and Antônio Olimpio de Santana of the Program to Combat Racism of the World Council of Churches – surprisingly, three men.

I have two daughters who are now 20 and 16, and a boy of 13. Moína (whose name means 'what shows'), is already married and has a 2-year-old child, my grandson Luan. Moína works as a typist and computer operator in the GRUMIN (Women's Group for Indigenous Education). My second daughter, Tajira (which means 'land's daughter' in our Indigenous language) is a champion in the first level of bodyboarding; my son Pontiguara likes skating, surfing and football.

My daughters are intelligent and are fully aware of the history of the women in the family. They are independent, with strong characters and can sometimes be authoritative. They had a financially unstable childhood and missed the tenderness of a father: that has been a problem for them.

I have been working for years to maintain the economic status of the family. My personal motivation has been the testimony of the years that I spent with the strong women who raised me. That testimony has been converted into an enormous reserve of willpower that other women might have access – through what we are, what we want to be and what we have been – to a more objective consciousness; and, through the creation of GRUMIN, to the mobilization and organization of Indigenous women to value their political role in the family, the community and the country.

When she was very small, my grandmother learned with the old women how to plant mandioca; she learned how to plant, to pick and to replant. My grandmother was wise because she knew how to plant mandioca, that which unifies the Indigenous People of Brazil, one of the greatest alimentary cultures.

The dignity of women like her, here or in any other part of the world, is the fruit of struggle, of the growing political role of women in their home, their countries and the whole planet. That growing role of women forecasts a new world for a new conception of thought between men and women, because men learn from women and become more sensitive with them.

An engagement with the Indigenous culture

I am sometimes asked what it means to feel that you are Indigenous. This is not easy to explain, because it is indeed something which you live. In 1992 I went to an Indigenous Press Meeting in Mexico. In the course of that trip, I became very aware myself of what being Indigenous means to me. I had the opportunity to climb the pyramid of the sun; from there I felt Nature Tupa (Indigenous God), the Universe and the sound of the drums, something I had never felt before. Ten days previously, I had been in the Indigenous Area of Pontiguara, working with Indigenous women. I was physically ill, and was being put under political pressure by the federal police, by groups with particular economic interest, including Indian owners of big properties, and other opportunists who are capable of anything for an easy profit. Around that time there was also a moral defamation campaign against me because I had denounced the passing of the Indigenous lands to the economic and political powers of the region. Coming from that world of filthiness and insensitivity, I was very happy to be invited to Mexico. There I found spiritual forces, dialogue with my ancestors and with some Indigenous leaders, that gave me the strength to continue that hard and lonely fight. It was there that I wrote the following text.

Dignity

The most beautiful thing that we have inside ourselves is *dignity*. Even if we are being maltreated, there is no pain or sadness that the wind or the sea won't extinguish. The purest lesson from the ancients comes from wisdom, truth and love. It is beautiful to flourish despite the feelings imposed by power. It is beautiful to flourish among the hatred, the envy, the divisions, the lies and the trash of society. It is beautiful to smile and love when a waterfall of tears enfolds our souls. It is beautiful to be able to say yes and go ahead. It is beautiful to build from nothing, and to open doors. A dignified future waits for Indigenous peoples all over the world. Too many lives, cultures, traditions, religions, spiritualities and idioms have been destroyed. The Truth will come out, even if our teeth are taken out; the important thing is to go ahead, to eat crab with flour, dried fish with tapioca cake and *mandioca* (manioc). The important thing is to watch the sea and the sky; to worship the dead, the ancestors; to dream their dreams and to see them; to live with the *manias de cabôco* that in fact are the most sacred links of our people because they are links with the Earth and the Creator, our god Tupa. It is beautiful to wear the garments of the Tore and to feel honoured as if you were wearing the clothes of the kings and feel it as the most important expression of the relations between men, the Earth and God. The important thing is to feel the sacred and the Universe; to believe and to trust even if, the night before, our house or our body was violated. It is necessary to listen to the old people, to the sounds of the sea, of the wind and the sun. It is necessary to unite the Pontiguara people; that is why we ask Tupa to protect us and to bring an end to the secular suffering of our shrimp-eater people, the Pontiguara people. We appeal to the superior force, which our thoughts raise to the most sacred levels of the Pontiguara Indigenous spirituality, close to the old shamans abolished by the powerful but reborn like force, through the consciousness of the people. We beg our souls to raise themselves to the sacred sphere of human wisdom and to receive the irradiation of love, peace and knowledge into all Pontiguara hearts. Turning all thoughts of conflict and disagreement into one peaceful thought that will build up the Indigenous Unity, we can build a great union of energies supported by everybody who reads this commitment in order to assure the dignity of a people abandoned and condemned to extinction.

No, we cannot accept defeat. There are youngsters, there are children smiling, there is the sea, the sun and there is hope. There is spirituality. It is enough to set free the chains of racism and egoism, those enemies that separate our people. Let's open the door, let's go in. Our ancients are waiting for us to join the ceremony of the deep peace and the unbreakable light. Let's also pray for the warrior, friend and protector of the Pontiguaras on 29 September. A great memorial will be built in honour of the shamans, the ancients and the warriors, the everlasting guardians of the Earth and nature.

Come, my people, let us raise our thoughts to Tupa and open our hearts in a 'Prayer for the Liberty of the Indigenous Peoples', not only for the Pontiguara but also for the 300 million Indians that live on the Earth – and let's think of the wise sentence of the Xavante Chief Aniceto: 'The woman's word is as sacred as the Earth'.

How do you build dignity? Dignity is like work. Societies passed from one phase to the other through work: the Time of Work! People took roots from the earth, then they built pottery to cook the roots and perfected dishes with roots like 'mandioca'. Mandioca is a basic food for the Indigenous people of Brazil, in the 9,511,965 square kilometres of the Brazilian area where it was used by about 130 Indigenous peoples dispersed through several states. Mandioca unifies the Indigenous people of Brazil. Mandioca is dignity. Dignity is built step by step. We must water it every day, just like mandioca. Dignity passes through human character, through experience, pain and suffering as well as through happiness, learning and love. Dignity is a diamond. The woman that passes from generation to generation suffering the political, social and economic process of oppression is accumulating a vast experience of feelings, of human relations; she has been building dignity. Dignity is built on social, human, political and economic relations. Through dignity, it is possible to see cultural relations that may or may not be oppressive.

Unfortunately in our society dignity lives under threat. In the blink of an eye the oppressing political system tries to throw our dignity in the mud, and we are always insecure, in spite of our force, our history and our past. That is how I feel when I am in the middle of the crossfire, when I am in the Indigenous village of Pontiguara, in the northwest of Brazil. There I find the oppressed and the oppressors. The oppressors are the philosophies and unagreed thoughts, oppressive, opportunist and cruel. Unfortunately, those thoughts have both white and Indian faces. The oppressed are the people who want to change but have to remain silent in front of the reactionary forces. That is the fruit of the colonization of the past and of the neo-colonization of the present, and a lot of people don't want to be aware of that monster. A thousand efforts are being made to raise the consciousness of Indigenous people, that bring with them the marks of a selfish, capitalist and oppressive colonization and neo-colonization. Power keeps creating other local powers, obtained by the rental of Indigenous land to groups with political and economic interests.

Who suffers from all of this? The women, the children, the old and even the many men of conscience who are capable of crying with us and who wish for social change. The International Indigenous Rights that have been worked on by a number of serious people in the United Nations don't arrive at this end of the world. Here those rights are violated, here the killing, the rapes, the prostitution go on; here everybody's dignity ends up in the mud.

When I am among lawyers or in big international meetings, walking over red carpets, the world seems beautiful to me. One writes things down very easily. The speeches are exciting. Everything is very easy. When I am in the bush, in the interior of Brazil, where information and development do not arrive, I find that Nature, although it is beautiful, is also the stage for treason, schemes and lies that favour economic power. The Environment Conference of Rio-92 glorified the environment but did not call attention to the hunger and misery in which the people of the rural areas and great forests live. It was silent about the great conflicts, sometimes armed, which are hidden from the press and the TV companies. In those conflicts the ones who suffer are the ill and hungry populations that end up marginalized in their society or exiled in their own country.

153

Hunger leads to madness, and madness leads to violence; destruction to suicide and homicide. To preserve the environment is to preserve the men and women of the planet above everything. To preserve them with work, culture, development, health and education. The Brazilian people are hungry. Hungry for food, hungry for education and health. The dignity of the people is in the mud.

I used to be afraid to write, but now, 45 years old, woman, mother and grandmother, I am not afraid any more. I understand why I was suffering from political persecution. What can I lose for telling the truth? I can lose everything except my dignity: that I will pass on to my children, my grand-children and my great-grandchildren. This dignity was built not individually but collectively, with the history, the ways of living, and centuries of experiences and oppressions of a people condemned to extinction: the Indigenous People of Brazil. That dignity was built mainly by the history of the Indigenous women of my country. Brazil was built by the black and Indigenous women.

12

LEFT TO INVENT THE FUTURE

Dolores Rojas Rubio, Mexico

Marina, 1915–

Luisa, 1943–

Dolores, 1963–

GRANDMOTHER, 1915–

Where our story begins

My grandmother was born in 1915. She spent her childhood and teenage years in the harbour town of Tampico, Tamaulipas, in the north of the Republic of Mexico. As in harbour towns throughout the world, life was centred around the boats and their comings and goings. One of my grandmother's favourite activities was to wait for the boats coming from far-away China, so that she could buy silk, lipstick, make-up and beauty products that could help her to look like Greta Garbo. As the daughter of a prosperous baker, she played the piano, played tennis and went to school. For my grandmother, thinking about her future meant thinking about meeting a husband who could ensure for her the same social status as her father.

My grandmother had the chance to study, and she took it. It would have been 'normal' for her to marry and for the marriage to be arranged by the parents of the couple. But my grandmother rebelled against this: she fell in love with her Accountancy teacher and, in spite of all the opposition, managed to keep both her lover and her studies. The only thing she agreed to give up was the Accountancy class which he was teaching.

She was convinced of the importance of taking care of her eight children and her husband. They were living in a small but growing industrial city: using the knowledge she had acquired from her own studies, she began to give typing classes at home, and persuaded her husband to teach her Accountancy and English. My grandmother thus joined the working world and began her career as a teacher. This is one of her proudest achievements: to have created an educational centre which still exists and to have gained recognition and acknowledgement for her work as teacher and head of the school. She is very conscious of the fact that she was the key person in conceiving this project and in bringing it to fruition.

MOTHER, 1943–

The pleasure of decision-making

My mother was raised in a small industrial city during a period that saw the building of a petroleum refinery which attracted people from neighbouring countries. During her youth – she is now in her fifties – it was common for the girls of the city to go to school, although anything beyond the basic range of studies was quite unusual, especially for women; for most, the first choice would have been Secretarial Studies. My mother went to school and studied to become a secretary in the school that my grandparents had started and where they were teachers. Besides school, sport was also important in my mother's life: she played basketball and softball, and was quite successful in both.

Although my grandparents had encouraged her to study, my mother's brothers were not very happy about her going out to work. In this period, the only women who worked were from poor families, where the men (father and brothers) could not earn enough to support the family. The ideal for a woman was to meet a man with whom she could build a family and a future.

But my mother made her own decisions. Sport was an important part of her life; she gained great satisfaction and self-confidence when others recognized her ability and called her '*chavalillo*' because she played as well as the boys. Against everybody's advice, she married young, a man who had neither looks nor fortune, and moved to Mexico City, where her husband could find better professional opportunities. Her married life thus began in an unknown city, far from her relatives and friends, with a man who was often away, travelling for his work. There, I think, my mother found energy in the challenge of the city. She very soon had children to care for (two in two years of marriage) and had to face sicknesses and worries alone. The image I have of my mother is that she could solve every problem, from repairing the house to fixing the car, as well as dealing with normal, everyday things and taking care of her children. I think that she got satisfaction and encouragement from the fact that she was able to do things that are more commonly done by men – and, moreover, that she often did them better.

I believe that my mother and my grandmother both experienced a feeling of completeness each time they took a decision. In the dominant life pattern, there were always some interstices left, some small areas which were open to decision-making – decisions which could be enjoyed, even if they ran contrary to the opinion of others. I recognize the influence of those two women in my life: I enjoy being a woman, being a single woman and living alone, having made the decision to face the challenge of discovering another way of living fully as a woman.

DOLORES, 1963–

I like what I do

I was born more than thirty years ago in the city of Mexico, one of the biggest cities in the world. Having spent the greater part of my life there, I can say that I am a city girl. This clearly determined my educational opportunities. In my youth, I was more or less obliged to think about going to university: the minimum education level of middle-class women of my generation was some preliminary study, and many of them completed a degree or even a doctorate. Although I had no choice about going to school, I went willingly and with commitment. I studied a subject which interests me greatly and which, according to my father, is a man's subject.

By the time they had finished their studies, a lot of my friends were married; some of them already had babies or were pregnant. Although most of these women feel some need (either personal or financial) to work outside the home, their wages are very seldom the principal income of the family. Care of the children is almost always the woman's job, so most prefer to work part time, at least while the children are young.

Among this generation, it is quite common for children between the ages of 3 and 5 to go to nursery; some babies are placed there before they are a year old. These little children hurry from home early in the morning with their mothers; they are taken to the nursery or the kindergarten where they spend the morning. If the child is lucky, there may be sport or art in the afternoons; if not, he or she will be taken to a grandmother's house, or to a friend or a relative who can look after the child for the rest of the day, while the mother is working. For entertainment, there will be cartoons on the television, or video games.

For most of these women, life has followed a familiar pattern, beginning with a period in their parents' house, going to the local school, aspiring to go to university, often going out to work while waiting to find a husband.

But what if life doesn't follow this pattern? It can happen. Differences within the family, especially between father and daughter, mean that some women leave the family home much earlier. This would have been a scandal just a generation ago; it is still not common, but it does happen. It happened to me. Suddenly I had to find work and a place to live, as well as continuing my studies. Because of my fascination for electronics, I was studying in the Faculty of Engineering. It was not easy to find work which left enough time for my course and paid enough to keep my stomach full. Simply learning to survive was not so difficult, but it was not enough: I needed so much more besides, like stability and security.

Sometimes I felt like giving up and going back home, but then I remembered the rows: 'Can I go to the cinema?' – 'Only to the four o'clock show, and you must come home as soon as it finishes'; 'Can I go to a party?' – 'Be back by ten!' Remember: 'Decent girls don't walk around the streets after ten o'clock at night.' I used to ask why my brother was allowed to wander around when he liked:

'Because he's a boy.' There came a time when I simply did not consult my parents any more. My father's final words still hurt: 'If there's a rotten apple it is better to get rid of it before the rest go bad.' I'm still not sure whether it was pride or anger, dignity or rage – or maybe a combination of all these things – that made me decide that it would be better, if I could manage, to live for life's sake, to ask my friends for help sometimes, if necessary, but to keep on searching until I found what I wanted.

In time I got to know a man, and fell in love. We lived together, formed a couple, got married. I continued doing my own things: I was beginning to discover myself and to find out what I liked to do. I realized that I was artistic, and felt a need to express this artistic side of my nature. This was when I decided to study Theatre as well as Engineering.

After a while I got a job as an engineer. In the evenings I rehearsed for plays with the theatre group. My daily routine was full and demanding, but I still had inner doubts. Things at work didn't always go well. Problems arose; people started to take positions, and there were difficulties with the workforce. The company asked for my opinion and I had to make a decision. I began to realize that there were many other things that were important to me – things I had never thought of before, like human rights. I began to get involved with a whole range of new activities which filled my time, my head and my life. Thus I arrived at politics, and my life as a militant began by working with an organization.

Then I turned around and saw my partner, distant from my new activities. He thought of them as a passing hobby which he 'allowed' me to pursue. I tried to share my interests, to involve him: I wanted him to participate. Mission impossible. He thought I was overreacting, exaggerating – that I should be getting over this 'phase' by now. He suggested that we start a family, so that I could stay at home and look after the children. Finally he became more aggressive: he didn't want a woman who spent all day on the streets.

I thought about it, but this was not what I wanted. I didn't want to be shut up at home, so why should I leave my work and give up my own interests? I didn't want to keep quiet about my ideas and my feelings. Finally I was forced to conclude that I didn't want to be with him any more. I was overcome by fear: What would I say to my family; how would I explain? Would they understand? I knew there would be pressure, people who would tell me that I was mad, that he was a good husband, that he let me do what I wanted; that he was handsome, earned good money; and if I had to do some of the things he asked, what did it really cost me?

In the end I had to make a decision: we got divorced. We both had to leave our rented flat – the owner didn't want a single woman there, much less a divorced one. The classic solution would be to go back to the parental home, but I rejected that idea. So it was a matter of looking for a new place to rent, facing all the usual excuses: 'Sorry but there are a lot of men living here and we don't want any problems'; 'Sorry but the neighbours are very particular, very decent', 'I don't mind myself, but there is always somebody who makes a scandal'. It isn't even possible to share with another single woman in case they think you're lesbian. So I had to do the rounds of friends and contacts once again, trying to find a group of people who would share a place and the costs. There had to be at least one man amongst us, to convince potential landlords that we were a decent group of people.

I felt strange, as though I had suddenly shaken off a heavy weight and was ready to fly. I was surprised at how accustomed I had become to blaming myself, at how anxious I felt about doing what I wanted to do and what I thought was right. I carried on with the theatre and the human rights group. I left the engineering company for a job in a publishing house, as a style corrector; I took some courses and then accepted the offer of a job outside the city. I was away for a year: when I got back, I picked up where I had left off in the publishing company and began a new relationship with a man. This time I really thought he was the right one for me, and fell head over heels in love with him. When it ended – terribly painfully, but I didn't understand why – I was convinced that I just had to keep going and learn from my experiences.

When a new work opportunity came along, I decided a change might do me good; I put in an application, had the interviews, and was offered the job. This was how I came to know another country and another reality: Guatemala. I worked on a project to commercialize artisanal products, with a group of people who wanted to help refugees. The team included many different nationalities: Guatemalans, Mexicans, Americans and Europeans. It gave me the chance to get to know Guatemala, its people, its customs, its languages, its world-view. More importantly, I got to know the women who produced the beautiful woven cloth which the group was trying to sell. I spent three-and-a-half years working with the project – years which proved decisive to my development and growth as an individual. I thought again about becoming part of a couple; I felt afraid, but wanted to make a serious attempt.

It seemed that 1994 would be an important year for my country, Mexico, so I arranged to be there for the whole year. I started working part time for Mexican non-governmental organizations (NGOs) and continued (part time) my work with the refugees. At that period I also felt that I was definitely in the best relationship that I had ever had.

And 1994 was indeed a very important year; through my own efforts I had made some space for myself in the political party, and participated in the electoral process. I found the political scene seductive. I saw, I wanted to learn. I listened to different ideas, different viewpoints, different positions. One thing stood out for me: all the women who had some kind of responsible position (and there were not many) had one thing in common. The women who were most active, who achieved the most, were single.

Now it was my turn. I decided that living with someone didn't work, and that I was prepared to be single for a long time – maybe even for ever. I decided to look for a different model for my life: I felt relieved to have made the decision, and wanted to create the circumstances to carry it off. From now on, I would go headlong for what I wanted.

A new opportunity arose for me, to support the party campaign in Oaxaca (a magical Indigenous state in the southeast of Mexico). I didn't give it much thought, I just went for three months: another indelible mark in my formation. My decision was confirmed: I really like politics. But, with the elections over, there was a general sense of frustration and demoralization; even more anger than before. What happens now? We are left to imagine the future, to invent it and defend it at the cost of hope.

None the less, I can say that I thoroughly enjoy my work and the chances it gives me to get to know different people and different places. I experience a sense of fulfilment each time I can say that I have done what I liked; or, at least, that I have liked what I have done.

13

AN IRISH MATRILINEAL STORY
A century of change
Ethel Crowley

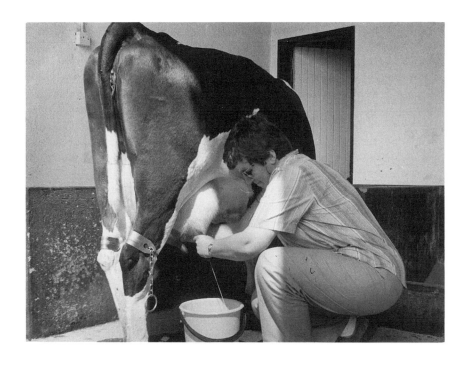

Mary, 1898–1994

Catherine, 1931–

Ethel, 1968–

MARY, 1898–1994

Mary was born near Bandon, a market town near the south coast of Ireland, in 1898. She spent her childhood living on a substantial farm, which her parents worked. There she stayed throughout her early years, working both on the farm and in the household. As was common at the time, Mary did not ever engage in paid work outside the home. She was a young woman of 22 during the Irish Civil War and indeed participated in the struggle for national independence. She was a member of Cumann na mBan, which was the women's wing of the Republican revolutionary movement. She married her husband, John, in 1926 and moved to the farm he had recently bought nearby. John, my grandfather had more experience of the world, having lived and worked in Boston, USA, for twelve years. He had been away from Ireland during the independence struggle and hence had no political involvement. It seems that he felt quite superior to his neighbours and friends because of his experience of America. His *savoir faire* and his wife's enterprising spirit accorded the family high status in the community. This relative sophistication earned him the nickname 'Old By Gosh', because this was an American phrase he often used. My grandmother's identity was forged very much within the context of the family she came from and the new one she formed. Mary lived a simple life in newly independent Ireland. Women of my grandmother's class worked extremely hard, both on the farm and in the home. Her social life would have been very limited, relying on the weekly trip to Mass and visiting other people's homes. She gave birth to seven children, which would have been slightly below average for that time, with no child mortalities, which were very common. Her husband died prematurely of cancer in 1951. Her youngest son took over the running of the farm and, when he got married in 1960, he sold the home place and bought a new farm about sixteen kilometres away. Mary went to live with him and his wife, where she stayed until her death in 1994. She spent her twilight years very happily there, where she was very much treasured. She had a very important role in rearing her son's children also, and was always at hand when needed. She died at the age of 96 and is fondly remembered by us all.

Context 1898–1930

My grandmother lived to witness many changes for Irish women and belonged to that group of Irish women who were 'written out' of Irish history until quite recently. Very few of the women of Cumann na mBan were active fighters, but more usually served the very important function of providing 'safe houses' for the Rebels and carrying artillery and messages to them. As is usually the case, however, once the war was over and the Republic of Ireland was formed in 1922, women were once again relegated to the private sphere and 'a budding feminist political consciousness was subsumed by nationalist hegemony' (Gardiner, 1993: 47). This period, when my grandmother was a young woman, was indeed one of the most active and fruitful times for women in twentieth-century Ireland. Women were enfranchised in 1922, after a long and bitter struggle by Irish suffragists over the previous fourteen years.

Domestic service was the most usual occupation for women in the paid work-force in the first three decades of the century. There were very few opportunities for women to escape lives of unadulterated drudgery, as is the case for many peasant peoples throughout the world today. The world that the suffragists fought for did not materialize for most. Most women, my grandmother included, would never have been encouraged to do anything else except to marry or join the nuns. (The dominance of the Catholic Church throughout the century is a theme to which I shall return later.) Her creative and participatory role in the nationalist movement was accorded very little importance. The search for integrity or individual creativeness and happiness was channelled either through marriage and motherhood or through a religious institution. These aspirations would not have been the primary considerations for young women at this time, by any means. Setting up their own home would have been the optimum goal of most. In the home, women's and men's roles were strictly defined according to the idea of their complementarity, based on their 'natural' functions.

There was a collective aspect to country life up until about the 1950s that worked like a system of mutual aid. For example, older people still have fond memories of the annual 'threshing', where all the neighbours would come to help bring in the harvest. Everybody worked, men, women and children and I'm sure nobody felt oppressed! When the work was done, everyone ate, drank and sang into the early hours of the morning.

Arranged marriages were common up until the 1950s in Ireland, particularly in rural areas. (To my knowledge, however, none of the marriages in my own family were arranged.) For many, this meant that the parents of the girl had the main decision-making power in deciding whom she would marry. Also, the girl was usually consulted to a small extent. This type of arrangement was in order to ensure continuation of the family lineage and also often amalgamation of farm holdings. In demographic terms, the inheritance system of primogeniture ensured that there was room only for one son, usually the first-born, to inherit the land and for daughters to marry 'into' another farm holding of comparable size and family of similar class. The 'excess' members of the family emigrated or worked on the farm in the capacity of a labourer.

There was then, and still is, a very strong cultural-symbolic strength attached to mothers and motherhood in Irish society and mentality. This is largely derived from Catholic teaching, where the mother is at once virginal and nurturing, pure and long-suffering. The relationship between the Irish mother and her sons, particularly, has traditionally been very strong. This type of thinking, which associates mothering with lifelong devotion to her children, is partly responsible for Irish women's low participation in the workplace and in politics, both of which I shall return to later. This predetermined role for Irish women is enshrined in the 1937 Constitution of Ireland in Article 41 on 'The Family':

> 1.1. The State recognises the Family as the natural, primary and fundamental unit group of Society, and as a moral institution possessing inalienable . . . rights antecedent and superior to all positive law . . .

> 2.1. In particular, the State recognises that by her life within the home, woman gives to the State a support without which the common good cannot be achieved.

> 2.2. The State shall, therefore, endeavour to ensure that mothers shall not be obliged by economic necessity to engage in labour to the neglect of their duties in the home.

Pope John Paul II, in his visit to Ireland in 1979, contributed to this construct with the following statement:

> May Irish mothers, young women and girls not listen to those who tell them that working at a secular job, succeeding in a secular profession, is more important than the vocation of giving life and caring for this life as a mother . . . I entrust this to Mary bright Sun of the Irish Race.

Not everybody believes this, of course, but this frame of reference is still of cultural import. The Irish women who have adhered to this model have indeed been praised in poetry and song, like the mother of Patrick Pearse, one of the national heroes of the Irish struggle for independence from Britain. Those women who have channelled their energies into other types of public activity have, at best, been written out of history until recently, and at worst, been labelled eccentric or unnatural.

CATHERINE, 1931–

The life experience of her daughter (my mother) bore a strong resemblance to that of her mother. Catherine was born in 1931 and relates many stories of a very happy childhood spent with her siblings. They lived on the farm which her father and mother owned. They all walked to school a few kilometres away every day, wearing boots in winter and barefoot in summer. There were always animals in and around the home, as often as not with a weak piglet being warmed in front of the fire and baby chickens scuttling about underfoot. There was always a lot of activity, led by the rhythm of the seasons. It was a mixed farm, with a little of everything. A few horses were always kept (my mother still retaining her deep love of horses) and, interestingly, flax was grown to service the (then) burgeoning linen industry in the north of Ireland. The children were given light tasks from an early age. These increased as the years went on, and by her teenage years Catherine was a full-time worker on the farm and in the home. She attended school up until the age of 12 and can still recall verbatim many of the songs and poems that she was taught there. She did not receive a secondary education, even though her brothers did. While her brothers were given the opportunity to learn trades, which were a pathway to upward social mobility, her role was already fixed.

In her late teens, she would go to dances accompanied by her brothers. A young girl was thought to 'need' a chaperone to keep her safe in such a setting. It was at one of these that she met my father, Denis. He lived a few kilometres away on another farm with his mother and sisters. After her father's death, they got married in February 1953, when Catherine was 22. My mother moved into his place. Her mother-in-law was a cantankerous and domineering woman, but she died six months after the marriage. Catherine had her first child in 1954 and bore six, finishing with me in 1968. My mother's search for integrity, therefore, was and is grounded in a small local context and filtered through her family and immediate neighbourhood. Her 'world' was a small local one, her opportunities limited. One of her outlets was attending the local meetings of the Irish Countrywomen's Association (ICA). I believe that the aspects of life that gave her something resembling wholeness or fulfilment were the real dignity derived from raising a family, working a farm, producing and reproducing. These are qualities of country people everywhere – the satisfaction in reaping a good harvest, feeding baby calves or eating a good dinner around the kitchen table with your young family should not require any further explanation.

Now, as a woman in her mid-sixties, she functions in an extremely independent manner, without a man by her side, as she was widowed nearly twenty-five years ago. She has managed her personal situation very well, but scorns how 'couple-centred' Irish society is, however, and how difficult people find it to deal with a strong, independent woman of her generation. This is almost expected of my generation, but not hers. She also enjoys witnessing the progress of her children, almost experiencing their joys (and woes) by proxy. So, according to established feminist criteria, her life has been lived basically for and through others, but in reality she has derived (and continues to derive) great pleasure from it.

She does now regret that she could not have had more varied opportunities in her life. She regrets that the many skills she possesses are not socially valued. She says 'my generation of women can do everything, and yet nothing'. I interpret this to mean that all the things she can do well as a result of her training are associated with homemaking and child-rearing, which are skills that are viewed as 'natural', i.e. unpaid.

Context 1930–1968

One of the most popular books in Ireland over the last three decades has been the Country Girls Trilogy by Edna O'Brien. This series of books tells the story of two girls from the rural west of Ireland who first go to Dublin, and eventually to England, in search of life and love. O'Brien was the victim of censorship from both Church and State in the 1950s and 1960s because she highlighted issues like sexuality and emigration for young women at the time. In her *Mother Ireland* (1976) she berates Ireland for being so oppressive to women. One could argue that she is particularly bitter, but a lot has changed since the 1950s, the period to which she refers.

Emigration has been, and continues to be, a salient feature of Irish life. It has been the main link between Ireland and the rest of the English-speaking world throughout the twentieth century (MacLaughlin, 1994). Analysis of this phenomenon has traditionally been located in the nineteenth century, after the famine (1840s), but the exodus continues. The 1950s were years of deep economic recession in Ireland and, being an economic barometer, emigration also increased dramatically. My mother's family was typical of the time – four of her five brothers left, two to the USA and two to England. None returned. It was more unusual for single young women to emigrate alone, even though it did occur, as in the cases of Edna O'Brien's heroines. As a result of this long history of emigration, I'm quite sure that there is no Irish family that does not have at least some close relations in England, the USA and Australia. If I can use such a term as 'the national psyche', I think that it extends beyond our borders to these places, as is evidenced in the Irish influence in music worldwide.

Rural life in Ireland up until the late 1960s was based very much on an extended-family model, with all that this implies. Help was always available on request from family and neighbours, and one of the only types of recreation available for a woman with a young family was to go on a 'scoraíocht', or a visit to family, friends and neighbours for an evening. This custom began to die out around 1970 because of the alternative types of entertainment provided by television and public houses (bars). Before then, outside of the major cities it would have been quite rare to find a woman drinking alcohol openly in a 'pub'. If they did, it was in a little alcove called the 'snug', which was used by a few female customers.

Discretion had to be maintained at all costs. The pub was very much the men's domain. Also, the national TV station, RTE, was set up in the early 1960s, signalling quite a change in Irish society. This period saw rapid liberalization of attitudes, when previously private matters were discussed in public for the first time. Neither was Ireland immune to the dramatic changes at the level of popular culture, for example, the miniskirt, rock and roll and Radio Luxembourg. Single young women then began to go out dancing, and many couples now in their forties or fifties would have met on the dancefloor!

Another form of recreation open to rural Irish (usually married) women, then and now, was attending the meetings of the ICA. This is a much underestimated organization which has been a very important social outlet for thousands of Irish women. It was formed in 1910 and its primary aim then was to provide 'a vibrant social life in the countryside as a bulwark against emigration' (Coulter, 1993: 33). This was not very successful, but it nevertheless was hugely important. It served the other important function of bringing Catholic and Protestant women together socially, serving to reverse the bigotry that lack of interaction and isolation between groups fosters. The ICA in the 1990s, while its ideology is not quite parallel to the dominant liberal, secular trend within feminism in Ireland and elsewhere in the West, continues to raise issues – like the treatment of women in Irish prisons and provision of health care for women – that might just otherwise go unnoticed. Perhaps it is an example of a type of 'Indigenous feminism' that is attractive to ordinary women and does not necessarily follow in the well-trodden path of Western liberal feminism. It is quite likely that its role has been, and continues to be, at least as important as (and perhaps more so) that of female representatives in government.

There occurred a domestic event about every four years that was vitally important for the integrity of country people, and especially the women. This was the 'station', which was, and still is, when the local priest came around to the house to say Mass, to collect the parish dues (money paid by the people to the priest for the spiritual upkeep of the parish) and to hear confessions. However, it was not as simple as it may sound. It involved, usually, redecorating the house and cooking a special meal for all the neighbours and friends. Afterwards, drinks were provided and songs were inevitably sung. The family, and particularly the *bean an tí*, the woman of the house, was judged according to how lavish the party was, rather than the piety of the occasion. It is commonly held that the Catholic Church introduced this custom, which still survives in rural areas, in order to instil in the Irish peasantry a pride in their homes. There was no guarantee that the house would be painted as regularly, in the absence of the station. Each farmer's wife [*sic*] did her utmost to host as good a station as she and her family could afford, and derived great pleasure from a job well done. It is a very mean-ingful indicator of wealth and/or generosity in the rural community. It was, and is, a popular social occasion where friends, family and neighbours gathered, but it unfortunately also had a very strong competitive element, which is hardly a characteristic of Christianity.

ETHEL, 1968–

I was born on my parents' farm in 1968, the youngest of six children. I was always used to being the centre of attention in a house where everybody was much older than I was. My father died suddenly when I was 4 years old, so my memory of him is very faint. My schooldays were very happy. I went to a small local mixed school where I could wear jeans as my 'uniform'. I was quite bright as a child, so the schoolwork was never a problem for me. Holidays were spent riding around on tractors with my older brothers, playing with the neighbours' children and with my ever-present dogs and cats. My secondary education occurred in Bandon, the nearest town, at the convent school. The nuns, however, did not have a high profile, hence the education I received was a relatively secular one. The academic standard at that school was fairly high. During my teenage years, I managed to balance doing well at school with having a good time outside it. I grew up quite quickly and, for example, had a steady boyfriend from quite a young age. Having started university in the nearby city of Cork, that relationship ended, as it coincided with his emigration. (We are still good friends ten years on.) As a student, I always spent my summers working in different places like Boston, Massachusetts, and London. After graduating with a BA at the age of 20, I went to live in London for a year, which was both alien-ating and exhilarating. I subsequently moved back to Cork in 1990 to begin an MA in the Sociology of Development. In this way, I indulged my insatiable curiosity about different places and cultures. At that time, I began a relationship with the wonderful man with whom I've been sharing my life since. Marriage is not on the agenda, as we both feel that it would not give us anything that we do not already have. We share a passion for travelling and we have spent time in several parts of the world, South and Central America being our favourite to date. I balance this with teaching in the Sociology Department at University College Cork and working on my doctorate. The topic of this is the effects of European Union (EU) agricultural policy on social structures in rural Ireland.

Context 1968–

Of the six children borne by my mother, three (daughters) received third-level education. This was made possible by changes in Irish society in the 1960s, when a financial support mechanism was instigated for third-level education for those who could not afford to pay. However, to gain the grant for university, a student had to do literally twice as well as those who could afford to pay. This had the effect of creating and maintaining a two-tier system within the university, as well as in secondary school and society in general. This situation changed about four years ago, but competition is fierce for places for everybody now. Despite the fact that university fees have been abolished, the type of secondary school one goes to, and whether one can afford private tuition, is still very important, repro-ducing class divisions. (It is too soon to assess fully the social effects of this policy move.) I isolate university education as an issue because I see it as the primary factor which differentiates my life from that of my mother. Personally, I found progressing through the educational system quite easy, largely due to the fact that I have two sisters who achieved very highly at third-level education. They have been very important to me in terms of providing role models. There were several factors of socialization which one has to overcome in Ireland, however. The main factor is a strong lack of confidence among young Irish people, and often the only way to overcome this is to leave the country.

Emigration is still with us. Of the six in my immediate family, four are away perhaps permanently. While living in other places and broadening one's experience is indeed positive, it is also true that it leads to serious problems in family relationships. The situation emerges where siblings have to rely on snippets of information relayed in occasional letters, photos and telephone calls in order to keep in contact. Divisions emerge between those who stay and those who go, between those who associate with an idealized view of Ireland from afar, a land frozen in time, and those who live in it. Parents suffer and, above all, mothers suffer. They sometimes regret having borne children in a country that is under-populated by European standards and yet somehow cannot provide a livelihood for its daughters and sons. New ideas about globalization of the economy and of popular culture tend to devalue home and nation, and tell us little about the social composition of the emigrant community (MacLaughlin, 1994: 36). The very real pain of separation is also obscured.

Since the 1960s, Ireland has witnessed many changes, especially for women. The ethos of the Catholic Church, since the foundation of the state, has been and continues to be intimately connected with State policy. This is especially the case with regard to issues that pertain particularly to women, e.g. abortion and divorce. A referendum resulted from a case where a man tried to prevent his girlfriend from going to England for an abortion. This caused such an uproar that the government was forced to intervene. This referendum in 1993 was very revealing of the Irish way of thinking: it requested a yes or no answer to three questions: whether one sanctioned the right to information on abortion, the right to travel outside the country for an abortion, and the right to abortion under certain circumstances (such as rape). The first two were ratified and the third refused. This means that people do not mind as long as our problems are exported. As it is, a conservative estimate of the number of abortions sought by Irish women in Britain is 5000 per year. In 1995, a referendum to allow divorce in Ireland was passed by a very thin margin, and was fully implemented in March 1997.

A real and symbolic example of the control the Catholic Church exerts over women's bodies and sexuality is the following: up until Vatican II in 1962, which was an institutionalized overhaul of the Church's thinking, women had to be 'churched', having given birth to a child. This was a blessing bestowed upon a woman by a priest to rescue her from her 'unclean' status (in which the Old Testament proclaimed her to be). This enabled her to re-enter the Church in a state of 'grace'. My mother was churched after her first five children, but not for me. She says that she didn't think about it at all at the time, that one just did what one was told. In retrospect, now she feels that churching was a very repressive custom that belongs in the Dark Ages.

The present generation of women owes a debt of gratitude to the feminist campaigners of the 1970s. Access to contraception was illegal then, and feminists like Nell McCafferty fought hard for liberalization of laws and attitudes. For example, in 1970 a group of Dublin-based feminists embarked on the train for Northern Ireland (where it was freely available) to buy up every form of contraception they could find and bring it back to Dublin. The police and the media were waiting for them on their return and their action had a strong popular impact. Legally, it was ignored until a few years later.

> It was the fervent hope of those who embarked on the Pill Train that the Irish government would arrest us on our return, making us instant martyrs and obliterating all our sins. If you want to progress socially in colonised Ireland . . . the first thing you have to do is go to jail.
>
> (McCafferty, 1984: 352)

This kind of radicalism, humour, imagination, cohesiveness and courage is very rare in the 1990s.

In my opinion, the Church is an extremely retrogressive force which has little or nothing to offer to women or men of my generation. Previous generations of women did not have the opportunity to question as mine does. I, apparently along with the majority of my generation, derive my integrity from rejecting it entirely. Most Irish people have a minimalist approach toward religious practice, but resort to the Church in times of need, or for the big family occasions like baptisms, weddings and funerals. The marriage rate in Ireland has dropped dramatically over the last fifteen years, with a corresponding sharp increase in births outside marriage, the rate of which is now 20 per cent. It appears that the institutional Church is losing its grip on the hearts and minds of young Irish people. The formalities of the Catholic faith bear little relation to the type of earthy, and earthly, spirituality that guides country people everywhere. This manifests itself in the daily miracles of the sun shining, crops growing and babies taking their first steps. This terrestrial grandeur is divorced from patriarchal rules and hero mythology, which is the stuff of Christianity.

The main source of integrity that was made available to a much greater proportion of women since the 1960s was the opportunity to join the paid workforce. The Irish government, from 1958 onwards, adopted an open-door policy for multinational investment. The types of work that this brought to Ireland were usually low-skilled, routinized and female. They also adopted a dispersalist policy of attracting the foreign firms to rural (often remote) areas. This had the effect of people entering the workforce for the first time, especially in the more rural regions of the west of Ireland. There were many problems with the types of employment instigated, but the overall effect was to give women a potential new source of independence, at least for a while.

Let us now turn to the employment profile of Irish women today. The overall female participation rate increased, for example, between 1984 and 1987 from 30.6 to 32.4 per cent. The majority of women workers are crowded into very few sectors. In the late 1980s, 77 per cent of women's employment was in the service sector, including teaching, nursing and banking/financial services. There has also been a significant increase in married women working part time (Blackwell, 1988). The issue of paid work is an important one, because it is often the first step for women into the public sphere whence issues that pertain to women can be raised. While patriarchy and all that it implies is indeed losing hold, women have to decide what is the best way to engage in this struggle. Some examples of legislative changes have been the lifting of the marriage bar from public sector jobs and the Equal Pay Act of the 1970s.

Women in Ireland are also seriously underrepresented in the corridors of power, either in occupational or political terms. The election of Mary Robinson as our President in 1990 could be interpreted as quite a coup for Irish women. Since her election, she has served as an excellent ambassador abroad and is indeed a source of great pride for many Irish people. She represents the liberal strand in Irish thinking, coming as she does from a legal background with a special interest in human rights and feminist issues. She is also a partner in a 'mixed' marriage (Catholic and Protestant). She should not be seen, however, to represent mainstream thinking in Ireland. Her role is a symbolic one only, where she has to prioritize the policies of government over her own personal politics.[1] There are only a handful of women in the Irish parliament. Whether or not they actually represent Irish women as a group is entirely different, but at least they serve as role models for young women, who may be encouraged to enter a public role as a result. This type of representation has offered women little. The women who gained office happen to belong to political parties which have an essentially conservative ethos. *Too* much change certainly would not be welcomed by them. Change has been very slow for marginalized groups in Irish society and it cannot be said that female politicians have contributed any more to this process than male politicians have. Most of the more positive moves have been instigated by the EU – for example, equality legislation and protection for part-time workers. So, while politicians make their promises and play their party games, ideological and economic subordination continues under the aegis of capitalism.

Community politics are now also becoming a popular means of 'empowering' women. This is exemplified by the New Opportunities for Women (NOW) programme, initiated by the present government. This is a policy of funding women's community groups, which is perceived as an effective policy for solving the problems of disadvantaged areas. There is a proliferation of these throughout the country, and they are indeed becoming more vocal on issues pertaining both to women and their communities. Mary Robinson strongly supports these initiatives. These certainly are a means of dealing constructively and democratically with social problems and certainly are a new source of integrity for thousands of Irish women. The NOW programme has a wider

173

social and class base than other groups, like the ICA, being equally strong in urban and rural areas. There is an inherent recognition of the need for dignified employment in this programme, whereas the ICA serves mainly to enhance countrywomen's social life. Poor and working-class women's voices are now being heard almost for the first time. Again, however, we need to err on the side of caution with regard to women's community politics. There is a common perception that community politics is 'soft' politics, while the 'hard' politics is left to the men. Women who are involved in these groups need to address these issues in time and not allow the two forms of politics to become a further embodiment of the public–private divide. This is quite an exciting time for Irish women, which makes it all the more important to deal with any problems as they occur, rather than to ignore them.

The variety of contributions to society that these women make is a testament to their creativity and flexibility. Coulter argues that these groups are part of a long tradition of resistance and struggles for survival among Irish women, and not just a product of 'second wave' feminism (Coulter, 1993: 48). Incidentally, Catholic nuns are often to the forefront of quite radical local initiatives, and a nun, Sister Stanislaus Kennedy, has written one of the few good sourcebooks on Irish poverty.

On a lighter note, Irish women are making their voices heard on the music scene. Probably the best-selling album of the last few years in Ireland has been *A Woman's Heart*, a collection of Irish women artists. In it is displayed a wealth of talent in terms of women's singing and musical performance. A spirit has thus been unleashed in public view that has been kept under wraps for centuries. There are very few examples of revered women performers from previous generations: Irish women have traditionally performed in private, typically at the fireside. Perhaps this difference between women and men is universal, as Tannen (1992) also speaks of this phenomenon of men dominating the sphere of joke- and storytelling. *A Woman's Heart* has ensured that there is a very lively and vibrant music scene where women's voices are welcomed and acclaimed, which must be very encouraging for the next generation of female performers. It certainly constitutes a challenge to the idea that the 'proper' place for Irish women is in the home, as in the time of our mothers.

Ireland also has a strong modern tradition of excellent women's writing. Kate O'Brien, for example, who wrote in the 1920s, wrote from a lesbian perspective before the term was even invented. The characters in her novels, such as *Mary Lavelle* (O'Brien, 1984), made readers think about women's experience in a way that would have been very challenging in real life. Poets such as Eavan Boland and Rita Ann Higgins also explore women's sexuality, feelings and life experiences in a very evocative manner.

REFLECTIONS

The writing of this three-generation story has been quite a unique experience for me. The methodology involved in contributing to this unconventional book was one of beginning with my own family experience, then comparing and contrasting this with those of other women from all over the world. I was liberated from the normal academic criteria of remaining detached from my subject, of providing hard evidence for my conclusions and of using impassive language in relating social patterns and trends. Here, I was encouraged to use the heart as well as the head to elucidate the meaningful changes in Irish women's daily lives. I was thus freed from the constraints of rigid models and ideas of 'objectivity', and requested to explain how people *do* live their lives and derive pleasure from them in the face of some repressive institutions and social arrangements. Emotion was allowed, as were bias and using one's own family as 'evidence'. My account of changes for Irish women was therefore an overtly personalized one.

In looking at the ways in which Irish women have lived their realities over the last three generations, it is essential to keep in mind the differential impacts of certain societal changes upon different groups of women. Women *per se* cannot be viewed as an homogenous group. Overall change has been slow, but particularly slow if one comes from an urban working-class or rural small farming background. Even one's access to different coping mechanisms was and is dependent upon one's position in society. In many cases, women and men of the same class background have more in common than women from different class backgrounds.

By focusing on one set of individuals like this, one can counter the tendency in feminism to generalize and depersonalize. Women's experience has been homogenized in much feminist theorising, based on a universalist construction of womanhood. Some non-Western feminists have criticized this tendency and I find their ideas very applicable to the Irish case. Gender cannot be isolated from other aspects of life like class, caste, race and nationality – reality is not manifested only in terms of gender. We cannot presume that all women are equally powerless, exploited and oppressed *because* they are women. If, for example, my mother did not *feel* oppressed throughout her life, does that mean that she wasn't? Can one 'colonize' her life experience and define it as something that she did not perceive it to be? It appears to me that social theorists need to listen to people more, to find out the meaning that they attach to their lives rather than imposing glib categories. Conversely, we can counter the opposite tendency of complete relativism, without adherence to *any* set of guidelines as to what a life should consist of in a particular cultural context. It occurs to me that feminist theory needs to allow for the personal power and adaptability of individuals to life circumstances, and indeed celebrate the many ways that women overcome limitations that are often imposed upon them.

Another very important aspect of change is just what we are hoping to achieve here – interchange of ideas, observations, hopes and dreams across frontiers with the aim of building a strong movement for change. It is an example of a proactive contribution to contemporary feminist debates, which attempts to forge unity out of diversity in a very real way. The women whose stories are related in this book have been contributing to their society in their own ways and have sought dignity and fulfilment, sometimes in the face of great odds. The precise issues might be different in each place, but the resilience and strength of women is the same throughout the world.

NOTE

1 In April 1995, as I write, Mary Robinson is at the centre of a controversy concerning the precise role of the president, and how much autonomy she can exercise. On a recent visit to South America, she shook hands with General Pinochet of Chile, despite having publicly campaigned against his reign in the 1970s. She has said since that she was representing the Irish people, and not acting as an individual.

REFERENCES

Blackwell, J. (1988) *Women in the Irish Labour Force*, Dublin: Employment Equality Agency.

Coulter, C. (1993) *The Hidden Tradition: Feminism, Women and Nationalism in Ireland*, Cork: Cork University Press.

Gardiner, F. (1993) 'Political Interest and Participation of Irish Women 1922–1992: The Unfinished Revolution', in A. Smyth (ed.) *Irish Women's Studies Reader*, Dublin: Attic Press.

McCafferty, N. (1984) 'Ireland(s): Coping with the Womb and the Border', in R. Morgan (ed.) *Sisterhood is Global*, Harmondsworth: Penguin, pp. 350–355.

MacLaughlin, J. (1994) *Ireland: The Emigrant Nursery and the World Economy*, Cork: Cork University Press.

O'Brien, E. (1976) *Mother Ireland*, Harmondsworth: Penguin.

O'Brien, K. (1984) *Mary Lavelle*, London: Virago.

Tannen, D. (1992) *You Just Don't Understand: Women and Men in Conversation*, London: Virago.

14

SUBTLETY IN PERCEPTION

Nicole Note, Belgium

Clara, 1900–1942

Alice, 1933–

Maria, 1959–

CLARA, 1900–1942

How would she have described herself?

Clara was born in 1900 in a little hamlet in Flanders, the northern part of Belgium which from medieval times was distinguished from the southern part (Wallonia) by language and culture. Today, Flemish is spoken in Flanders, while French is the official language in Wallonia. In the early days of independence (1830), however, French was the official language throughout the country and Flemish was considered the language of the lower class. It was only in 1930 that Flemish became the official language in Flanders.

Clara's parents were independent farmers with a relatively good lower-middle class income. They owned four cows, two pigs, two sheep, a horse and a dozen chickens, as well as beehives. Like all farmers of that time, the family also had a small plot of land where they grew flax and linseed. During winter evenings, when there was little to be done on the farm, the flax would be worked on. This was all done by hand, since most families had neither electricity nor machinery. The flax could be sold to spin linen for clothing, while meal or oil could be made from the linseed.

Clara's family was deeply Catholic and God was part of their everyday lives. Out of a combination of respect and fear, the family tried to behave according to his laws. Together they would thank him for their meals, make a cross on the bread before eating it and carry candles during thunderstorms; at night, before sleeping, each person would kneel to ask forgiveness for her/his daily sins. According to Clara's daughters, Clara did not at the time feel this to be a 'dominant framework' in her life, as it might be described today. Rather, it gave a

178

meaning to her life; within this framework and these boundaries she was able to live her life with considerable flexibility.

Clara was the third of nine children. Although compulsory education was not introduced in Belgium until 1914,[1] she and her sisters and brothers went to school up to the age of 12. The school system was such that during harvest time children could stay at home to help the family on the farm. After primary school, Clara wanted to go on to learn the profession of seamstress. Her parents agreed to this, and at the age of 16 she set up her own workshop. Her sisters helped her during wintertime, and she had room for three pupils, to whom she taught the profession. Her specialization was sewing the garments used for dead bodies. Clara provided a stable extra income for the family until she married.

At the age of 26 Clara married Raymond, the son of a flax farmer. Since they were related, they had to ask permission from the Pope for the marriage. Clara kept on her profession as seamstress, bringing in a good income, while Raymond continued as a flax farmer. Soon after marrying, they had their first child, a boy named Staf; three years later came a daughter, Paula, and another three years later the last child, a girl named Alice, came into the world. Clara kept on working, with two girls helping her: it was quite natural to her to combine work and children. She was usually cheerful and often singing, making her own sewing patterns. Man and wife both enjoyed their jobs. Raymond had two farmhands helping him with the flax. By chance, he was able to receive electricity on the farm, and managed to buy machinery to process the flax, both of which saved him a lot of hand labour.

Raymond and Clara knew all the people in their village and, to a lesser extent, those in the adjacent villages. The villagers depended on each other for some of the farm tasks such as harvesting. In the summer, the neighbours would come over frequently, sitting in front of the house to talk, to sing, to gossip. This solid social control system clearly limited individual freedom and creativity, but it did provide security. In times of trouble, there would always be emotional and financial support. Moreover, the social control aspect of the community tempered negative forces in the individual: according to Clara's daughters, women felt quite safe to walk home alone and at night (apart from during the war).

Raymond and Clara had a truly loving relationship: their daughters do not remember their parents ever having arguments. On winter evenings, when Raymond had finished work and Clara was sitting sewing, he would read aloud to her from books and Catholic newspapers (others were not allowed by the Church). They seemed to live very much in harmony.

The prosperity and happiness that characterized the beginning of their marriage turned in the middle of their lives. With the advent of the Second World War, the family was forced to flee and live with three other families in a barn, where they survived for ten days on nothing but milk. Belgium soon capitulated, however, and those who had fled were able to return to their homes. Paula, the older daughter, remembers the trip home very clearly: there were dead bodies everywhere, and German soldiers carrying guns. They were not unfriendly, however: when one of the family's cows escaped and Clara asked a German soldier to go after it, he obligingly did so, and brought the cow back, smiling. Clara then explained to her daughters that things had been different during the First World War: then there had been serious fighting between the Belgians and the Germans, and the latter had been much more hostile. Now that Belgium had capitulated they were not considered a threat.

For the rest of the war, the family had to share their house with German soldiers. The Germans were disciplined and friendly, but the family hated them and were scared of them none the less. The family had no privacy or freedom. Soldiers would interrupt them while they were eating, demanding some of the scarce, secretly prepared food: there was no way of refusing, or of hiding it. At all times of the day, the soldiers would drop in – for four long years. Compared to other villagers they were lucky: in wintertime Clara was able to make two small coats for her daughters out of her own coat; Paula learned to spin with the wool from their own sheep, making socks and warm sweaters.

Before the war had ended, tragedy struck. Clara suffered an attack of appendicitis; she went to a clinic where she underwent an operation, but the appendix ruptured and toxins escaped into her blood. For ten months she was seriously ill and bedridden, undergoing various operations. She finally died at home, at the age of 42. Just two years later, after the war was over, antibiotics were introduced in Flanders.

At the end of the war, Raymond was left with three children and little money. He had not been able to work during the four years of war: the hospital treatment and funeral of his wife had been expensive. On top of that, the government decided on a financial reorganization after the war, changing the currency. One of their motives for this was to thwart those who had made ill-gotten gains from the war: people who could not prove where their money came from received only a fixed minimum amount of the new currency. As Clara and Raymond had kept their money in a sock under their mattress instead of in the bank, a lot of money was lost.

Raymond took up flax-processing again, this time also renting land in Wallonia. Staf and Paula had to stay with a local family for up to a fortnight at a time, their father joining them occasionally. In Wallonia the war had destroyed extensive areas of forest and the land was cleared for agricultural purposes. Since the people of Wallonia did not know how to produce flax, and had neither the machinery nor the potential market, this seemed a good opportunity for Raymond. However, the soil conditions were such that the flax was of lower quality and fetched a lower price. Then, in the 1950s, a machine was introduced by one of the richer farmers which could process flax on a large scale. Prices went down; small farmers needed twice as much land to earn the same amount of money. Around the same time, nylon invaded the market, gradually replacing flax. Raymond gave up flax-cultivation and started selling beer, refreshments and coal instead. This went well, and when Raymond retired Staf took over the business. Because of the expansion of supermarkets in the 1970s, however, the business collapsed and Staf ended up poor.

Raymond lived with his son, daughter-in-law and granddaughter until he died of a heart attack at the age of 73. His two daughters were better off than their brother: Paula kept up the family tradition and married a good middle-class farmer. The couple had three children. Paula's husband died three years ago, at the age of 60, of cancer. Paula still finds it difficult being alone in the evenings. Alice, Clara and Raymond's younger daughter, married an army officer.

ALICE, 1933–

Taking life as it comes

Alice was Clara and Raymond's youngest child. She was 9 years old when her mother died, and was raised by her father, whom she admired and respected right up to his death, and by an unmarried aunt, of whom Alice does not have happy memories. For her, it was not this aunt but her godmother who partly took the place of her mother. She remembers coming back from church on cold snowy winter mornings, with an empty stomach – one was not supposed to eat before church – and passing by her godmother's house, where warm pudding and a gentle, welcoming atmosphere were waiting. Alice hardly remembers the war: according to her older sister this is because she was small and weak and, since she was the youngest, protected. Alice does not remember having been weak, although she frequently fainted in church.

Paula took on most of the responsibility in the household, but Alice also had to work hard on the farm. When Paula and Staf were in the southern part of Belgium working on the flax, Alice had to run the household on her own; and when her godmother died a few years later, Alice and Paula were enrolled in taking care of her young children.

Notwithstanding the work, Alice managed to study domestic science up to the age of 15 (although education was only compulsory to the age of 12). Her father and the local priest encouraged her to continue studying and to become a teacher. However, although Paula never asked her to do so, Alice refused because she felt she needed to help her sister with the household chores. She stresses that it was her own decision, her own feeling of responsibility that made her refuse that opportunity. Looking back at her life, she now thinks that she should have taken the chance to study.

At the age of eighteen Alice started to work in a shop. At first she was selling biscuits, but she did so well that the owner transferred her to selling women's underwear. His profits increased considerably while Alice was there, and Alice herself has happy memories of her work. In fact, looking back on her youth in general, she remembers it as a very happy and joyful period – a time in which she was free to be young and to do what pleased her, and in which she got plenty of encouragement to find herself.

At 23, Alice married Louis, who was 25. Louis was the son of a shoemaker, an intelligent and humorous man always in for jokes and tricks. Since his parents could not afford to send him to university, he joined the army in the hope of continuing his studies that way. He served his country by removing dangerous unexploded bombs from the two world wars. As a result he does not hear very well any more (although his children sometimes suspect him of only hearing what he wants to hear).

Because of Louis's job, the couple had to move to the French-speaking part of Belgium soon after their marriage. Alice had to stop working in order to join her husband. Here in the south they had their first daughter, and fifteen months later a second girl was born. Immediately after her birth, the family was transferred again; this time to Germany. This was a difficult step for Alice: there were no motorways as there are today, and they only had a very small car; phoning was expensive and it was not the custom to have a private phone. So for Alice this move meant being even further from home and far from her beloved father. To this day, Alice feels that she failed in her love towards her father.

Furthermore, her husband turned out to have occasional bouts of bad temper of which she was afraid at the time. Since she was not acquainted with this sort of behaviour, she did not know how to handle it. For the sake of the children, she tried to create an atmosphere of harmony and nearly always gave in to him. She says of those days, 'I stood there in between the children and my husband. But nobody ever asked me what I wanted.' She thinks that one reason for not fighting back was that women of her generation were told to submit to their husbands, and she did not know any better. She accepted life as it was.

None the less, she felt generally happy and carefree in this period of her life. Because of the spontaneous personalities of both Louis and Alice, they developed lifelong friendships with their neighbours in Germany. Once a week Alice would go to the cinema with the two other women, while the men played card games and looked after the eight young children (Alice gave birth to a third daughter in Germany).

An additional factor in making this a relatively carefree period was probably the economy: this was the beginning of the so-called 'golden 1960s', when mass production flourished, luxury products were easy to come by and were available even to the working class. A general optimism prevailed, with the belief in unlimited possibilities for economic growth and the further prosperity that would generate. Alice and Louis felt and shared that optimism.

After ten years in Germany, during which they had lived in a number of different places, Louis was transferred back to Belgium, close to Brussels. In Germany, housing had been free for the military: in Belgium they had to rent a house and buy furniture. This meant a serious budget readjustment. However, Alice was very shrewd in managing the household money and Louis trusted her fully in that. The children were not aware of any differences.

Alice very soon made new friends, but to integrate herself further into the village, she joined the Catholic women's movement. In the 1960s and 1970s the Women's Movement became quite strong in Belgium, although it was not comparable to the radical movement of The Netherlands, where women very much stressed individual independence and equality with men. The Catholic women's movement interpreted emancipation in its own way. It offered educational activities with regard to women's rights, general courses to understand political life, as well as typical 'women's' courses. It also provided a place for social contact and friendships. Alice joined for both reasons. She felt the need to expand her mental horizons: she chose the right moment, because times had changed.

In the 1970s the oil crisis showed the limits of economic progress. The first cracks in the general optimism appeared, insecurity about the future slipped in quietly. Ongoing scientific innovations and economic prosperity had encouraged people to think that everything ought to be possible. Alice felt this keenly as her children became older. One of her daughters started to take the pill at the age of 18, another went to live with a boyfriend. From within the relatively loose but meaningful (religious) ethical framework which Alice had inherited, these acts were clearly perceived as 'wrong'. At first, she held on to her framework and rejected her daughters. Later, however, she was able to find a way to adjust and to reorient her thinking, gradually taking a more critical stance towards statements of the Pope or the priest, tailoring and individualizing her ethical framework. During this transition period, she avoided talking to other people about these issues.

Although they caused some problems as teenagers, her children have always been important in her life. They served as a goal to live for, and she devoted many years of her life almost exclusively to them. Once they started to grow up, Alice had more time to herself, and started giving confirmation classes. It was then that she realized how much she liked spending time with children, and how she would have enjoyed being a teacher.

Since her children left home – initially a very difficult time for her – Alice has joined a group of women who visit the elderly and the sick in the village, and has started helping the nurse who checks young babies once a month – all on a voluntary basis. Because of these activities and her very open, caring and joyful personality, she is genuinely loved by the whole village.

When analysing her own life, Alice sees two main driving forces: social contacts and her children. She feels that if she had had financial independence, she would have been better able to stand up to her husband at crucial moments. However, she has managed to live an independent social life. Today, she tries to do what she likes, searching for a balance between her desire to socialize, her wish to help her own children with their children, and creating harmony at home. She feels quite happy with her general situation.

MARIA, 1959–

Following inner motives

Maria was born in Flanders, in the same village as her mother. She had one older sister and one younger sister. Up to the age of 12, she lived in Germany;

then the family moved to a village close to Brussels, about a hundred kilometres from the village in Flanders.

While in Germany, she had a very carefree childhood. The family lived in a village set up for Belgian military personnel, where everyone knew each other. School was close by and was run by the Belgian authorities. Life was easygoing and fun. Although her parents came from a Catholic background, they never pushed Catholicism on to their children: they had to go to church, of course, but it was not a family custom to pray at dinner or in the evening. Nor was this the case at the State school she attended. The children of the village were very free; their parents trusted them, and they spent a lot of time playing outside, indulging their fantasies with many other children of their age, girls and boys together.

Coming to Belgium disrupted this happy, carefree life; going to a Catholic school was a shock. Although the teachers no longer talked in terms of 'sin', there were many things one 'ought' not to do: for Maria, behaviour that had been pure and normal, like having dates with boys after school, now seemed to be contaminated. To Maria, Catholicism thus came to have a very negative connotation, and throughout adolescence she deliberately behaved in ways that the Church did not approve of, such as living together with a boyfriend. In fact, however, she only succeeded in shocking her parents with this. The environment in which she now lived was such that one could express individualistic ideas without being sanctioned. Of course, if they had lived in the small country village in Flanders, the situation would have been more uncomfortable for her than in the urban area around Brussels, although it would not have been impossible.

During adolescence it became increasingly clear that Maria was a sensitive person. She began to perceive reality very negatively at that time and developed an inferiority complex. There was, however, one constant factor through which she seemed to gain the strength to continue with life: by searching for silence in nature. She could find her balance again by walking through the majestical forest close to her home, finding herself while sitting among the high ferns in places where human beings were not, in fact, supposed to come. Weekends and long vacations in the countryside of Flanders provided longer periods of relief. Sitting in the soft grass at noon, with only the sound of the larks, the heavy heat embracing her, merging with the earth surrounding her – this offered her inner peace. She had experienced the same feeling of harmony at her godmother's farm. There she would spend hours on her hands and knees, searching for fresh potatoes in the dry earth, the sun shining down on her back. Even now, the feeling of being absorbed by nature gives her strength. (Later, a motorway was built through 'her' area of silence. Her own logic of justice was incompatible with adult rationality. It was her first questioning of 'progress'.)

During Maria's late adolescence, the 'golden 1960s' of unlimited opportunities were followed by an era of increasing doubt about the future. A growing number of people became aware that their way of life was damaging such basic necessities as air, water and earth, ultimately threatening their own survival. A

'Green' movement slowly developed and entered the political scene. Unlike the rest of her family, Maria was quite open to this alternative way of living and became converted to a macrobiotic lifestyle. Living consciously with the 'green' ideal in daily life, she felt herself growing and becoming more complete. Today, Maria is a member of a local pressure group, trying to preserve for the next generation whatever nature remains unspoilt. She considered going into politics, but withdrew, although it is not clear (even to herself) why she made this decision.

Without doubting the importance of the Green movement, transferring 'nature' into the science of ecology had its negative sides as well. Until the arrival of Green politics, the only people broadcasting apocalyptic messages of the 'world will perish' variety were fanatical religious groups: from the 1980s on, scientists gradually took over this job. Frightening visions of overpollution, lack of resources, etc. may well be one reason why people of Maria's age have fewer children (there was a decline in birth rate in the 1980s): it undoubtedly influenced Maria's decision.

When she was 21, Maria met her husband Wolfgang, a psychologist working on pain research. After a one-year trip to America they moved for his job to a small village in the east of Belgium, a hundred or so kilometres from the family, in an area with (by Belgian norms) still a considerable amount of relatively unspoilt nature. Thanks to easy transport and communication, Maria has been able to keep in close contact with family and friends.

Maria sees Wolfgang as a wise and moderate person; he is a buttress in her life. He has taught her to experience life with more joy. Because of his character, she has had to make few compromises: when she wanted to change career (she was an occupational therapist but did not feel satisfied in that role), Wolfgang not only agreed but even paid for her studies. Maria began to study Indian American Cultures and Languages at the University of Leiden (The Netherlands), which meant that she only came home at weekends, and was away in Peru for four months and later in Chile for six months. Wolfgang agreed to all this, probably because he felt that there was little choice. At that time, it was something that Maria needed to do.

Looking back at this period, Maria now recognizes that she was in fact searching for meaning in other, more nature-oriented societies, perceiving her own society as in essence ill-conceived. Through her experiences overseas, she could leave her black-and-white visions behind. Gradually she was able to see the complexity of her own society, as well as that of other societies. As a natural consequence, she also became aware of the absence of ways to value – and thus compare – these societies. She also discovered how the lexicon of development restricted her thinking about the issues. Her major goal now is to contribute to a usage of development that can express the complexity of developmental reality, going beyond existing stereotypes.

Four years ago Maria and Wolfgang had their first child. It changed Maria's life. She is now faced with the choices that many of the women of her generation have to make. Like a flower opening up, life has become richer but also more complicated: the more leaves and petals – the more aspects to life – the more friction may be created.

Maria feels that because of her children she experiences greater inner peace, as she can now 'place' herself in the world. She has rediscovered the importance of real relationships. Having children has also brought her closer to the other women in her family, and having a good relationship with the family has become important to her.

Looking around her and talking to other women, Maria has realized how important relationships are for many others, too. Thus what she had previously perceived as weakness and 'giving in' now often seems to her to be an act of loving. As one woman put it, 'I make compromises for the relationship, not for the sake of peace and quiet, but because the relationship is very important to me, more important than myself. I make compromises because I love my husband.' In extending this, Maria found that in practice quite a few female acquaintances let themselves be guided by many values and unwritten rules within their immediate surroundings. By listening carefully to them, however, it seems that they do not feel obliged to do this: they prefer to conform for various personal reasons, such as not wanting to spoil a relationship, or avoiding the unpleasantness which a certain subject provokes. They make these decisions on a relatively free basis, choosing to consider someone else's feelings and thoughts as well as their own.

Another important force in the lives of Maria and her female acquaintances is the education of their children. Many of them (including Maria) believe that the subtle guidance of children into behaving as 'decent' human beings is a fundamental contribution which they can make to society – although society itself may not value or recognize it as such. In concrete terms, this means spending time with the children. Thus, wanting to take care of children is often not an act of selflessness, sacrifice or compromise, but a well-considered choice. It is a question of priorities.

For most women, children offer a degree of fulfilment, but they are not the only source. Like Maria, many other women feel a desire, an eagerness or a determination to explore other aspects of themselves and of life. The most common way to achieve this is by working outside the home, not least because this is a choice of which society fully approves. However, stable, long-term jobs are increasingly hard to come by, as jobs in general become more scarce. So women creatively seek other ways out. One artist Maria knows formulated it as follows: 'The more I strive to come close to myself, the less I make choices on a rational basis to be happy. I learn to live by my feelings, to know what is important in life and what is not.'

The challenge now for Maria and these other women is to combine the different desires and parts of their lives into one balanced whole. They want the best of everything – for everyone involved, including themselves – but this is not always easy to achieve. The majority of these women have busy and demanding lives, running from one place to another. Their ability to create inner balance depends on different factors, such as the way the children behave on a given day, the pressure of work, their own psychological state – even the weather. The way they perceive their situation and respond to it thus depends on a particular context in which multiple factors play a role and can, within limits, change from day to day.

REFLECTIONS

I have tried to place the three generations here in a wider context, linking each of the women with a movement or spirit of her time. It is clear that people of my country have been prone to rapidly changing worlds, their own Flemish world criss-crossed by international movements. This has led to changes at the material and social level, as well as in ethical value systems.

Materially, there has been an enormous growth in consumption since the time Clara was born. From the late 1970s on, women have experienced both the consequences and the limits of this growth. With increasing mobility, women have gradually been able to extend their social horizons and their friendships; they could escape the negative side of social control. However, loneliness is a much more common feature today than in Clara's time.

Changes in the material and social sphere affected behaviour and ethical frameworks, and vice versa. The most important dynamic cultural change over the generations has been the change in the ethical value system through which the generations perceive(d) and live(d) their reality. During Clara's lifetime, the value system of the Flemish people was very much linked with an authority, the Catholic Church, that provided Clara and her contemporaries with a purposeful position in the universe and that told them how to live their lives. There were clear and severe divisions between good and bad. During Alice's life, technical innovations and mass communication have consistently challenged this authority. The credo became 'God is powerful, but so are men.' Women had to revise their guiding principles. But at least they could believe in ever-increasing welfare and hope that their children would have a better quality of life than they had experienced. This has changed again in the lifetime of Maria, as the negative consequences of unrestricted progress became clear. As a consequence, the earlier wave of optimism has turned into a wave of realism, and even of pessimism about the survival of the species. More and more people question the values of progress, if not progress itself. But no new values have developed from within society.

Today, the 'world-will-perish' messages continue in the context of the dangers of nuclear waste, overpopulation, etc., accompanied by a heightened economic instability. Young people build up their own value systems, influenced by their immediate environment, law, the media, and their own critical characteristics. They have more freedom of choice and movement, but they are also entirely responsible for it.

This leads to two methodological questions. The first is this: Can we scientifically prove that we have made steps forward? On what terms can we measure progress; do we take account only of the material side of life, or should intangibles such as happiness, inner peace and fulfilment be part of the equation? Which of our three women was the best off? – Clara, who lived within the limits of her village with its strong social control system, and her own narrow Catholic value system for guidance, but who had a loving husband and a fulfilling job? Or Maria, who could continue studying as long as she wanted, who was far better off materially, but who felt that she was living with the sword of Damocles hanging above her head? Is it even possible, in academic terms, to compare situations like these?

The second question relates to interpretation. In collecting data for this chapter, I found that Clara's daughters would disagree about the interpretation of some facts of her life. The same happened while comparing the stories of Alice herself with those of her daughters. Clearly there are different ways of perceiving the same reality. The daughters look at their mother's situation from their own perspective, starting from their own world of experiences with husband, family life and own personality. Thus within a given context it is important to know who is perceiving a certain situation (a Catholic framework, a social control system) as 'narrow' or 'strong': The person living it, or the person describing life stories? Extending this as the second main question, How can we ever be certain about our judgement of the quality of lives of others?

These questions are raised with a purpose, in the hope that they might be a contribution to a wider perspective, stressing the issue of subtlety in perception, and its importance for true knowledge and understanding.

NOTE

1 According to the Belgian constitution of 1830, women and men are legally considered equal. In practice, however, women have been discriminated against at the juridical, political, economic and educational level. The first (French) school for girls did not open until 1864; the first female student was not admitted to a (French) university until 1888. A turning point came with the First World War – during the war, women organized first aid, cooking and cleaning in hospitals, on a voluntary basis. This gave emancipation a new impetus, and in 1920 women were able to vote in local elections. In 1948 women took part for the first time in national elections. From 1985 onwards positive actions have been taken so as to promote women to top economic and political functions. In their jobs, women are well protected by law, including legal measures against sexual harassment by men (introduced in 1993). Legally, and in practice, women and men today are on an equal footing.

15

THE SWEET AND SOUR FRUITS OF WOMEN'S LIB

Edith Sizoo, The Netherlands

Willemijntje Sollman, 1875–1969

Truus Hoeksma, 1905–

Edith Sizoo, 1939–

Solange De Boer, 1964–

Manon De Boer, 1966–

WILLEMIJNTJE SOLLMAN, 1875–1969
'Money is a good servant, but a bad master'

Willemijntje was born in 1875 and lived for ninety-four years. She saw the first horse tram riding through Amsterdam and the first electric train in the country; she watched on television as the first man landed on the moon. After Willemijntje's birth her mother sent someone to get a pound of horse meat to help her regain her strength. Her father, coming home from work, welcomed the baby and went out to buy his wife a currant loaf, which he used to do each time a child was born. And life went back to normal.

Willemijn's father was the owner of a forge and had eight labourers working for him. He was a silent, withdrawn man: 'It is better to say nothing and that's all I have to say.' His voice was almost exclusively identified with Bible reading, which he did at the family table three times a day for twenty minutes, while his wife and the girls sat knitting and the boys kept silent.

Her mother, in contrast, was an outspoken woman with a strong personality who educated her children by serving as a model of the Calvinistic work ethic, reinforced with texts taken from the Bible and an apparently limitless reservoir of old Dutch sayings: 'As long as you are on your feet, your bottom won't mould.' She bore twelve children, five of whom did not survive childhood. For this recurring source of grief, Willemijn's mother found consolation in her faith and prayers. It was she (not her husband) who decided that one of her sons, who wanted to be an actor, should start a bicycle workshop instead. Although she had not approved of his wife, who was of lower social status, she forbade him to divorce her. She took care that her daughters stayed home to help in the household once they had finished primary school. Willemijn's mother died at the age of 95.

One of Willemijn's sisters was endowed with a gift of clairvoyance in the realms of human behaviour, birth and death; one of her brothers was also clair-voyant but in the medical sphere. He practised this all his adult life, without asking for money: 'shrouds have no pockets'.

In her early twenties Willemijn married Willem Hoeksma. Willem had lost his father at the age of 3. His mother, who was left without money, sent him to a Protestant pastor in Germany, who taught him at home until he was 16 years old (reading, writing, arithmetic, Latin and Greek). Willem was then sent back to The Netherlands and became a young office boy at the Standard Oil Company, later called Esso. He stayed there all his working life, ending as a member of the Board of Commissioners. Willem was a silent, self-taught man, who valued his privacy. The daily stock market news was his form of sacred literature and he enjoyed horse riding and going to the gentlemen's club. His weakness was his addiction to speculating on the stock market: unfortunately he always lost. Willemijn suffered a lot from the constant insecurity which resulted, although she supported his commitment to putting money aside for the Church, whatever the financial situation at home.

Willemijn and Willem had four daughters and one son. Although their first two daughters were not allowed to go to secondary school, this was not because their parents discriminated between boys and girls in terms of education: it was partly because there was not enough money and partly because there was no Christian secondary school in town. In later years, their son convinced them to allow their youngest daughter (a late baby in the marriage), to go to university to study Law, promising that he would take care of the finances. Willemijn and Willem were to be very proud of this daughter, who joined the resistance move-ment during the Second World War. After being imprisoned, she was the only one of her group to survive, defending her own court case before the Gestapo. When liberated from prison in May 1945, she met a Canadian army officer and left with him for Canada. Willemijn and Willem never went to see her. They had read a lot about Canada, but it never crossed their minds to go so far away just for a holiday.

Whether Willemijn really wanted the five children she bore is an unanswer-able question. Having children was not a matter of choice or discussion: it just happened. Her eldest daughter, now 94, still displays her frustration at the lack of love and attention she received as a child and at having regularly been sent away to Willemijn's sisters' houses. Her other children do not deny that their mother was not a cuddling type, but are somewhat milder in their judgements of her.

Willemijn loved reading the newspaper. Her favourite pages were those on church life and on national and international politics. In 1938 she suggested to her husband that they might spend the summer holidays in Germany, so that she could see for herself what this man Hitler was like. Having heard him speak at a mass meeting, her judgement was as simple as it was true: 'He is an indecent man.'

She was most interested in social problems. Although homosexuality was still considered a sin in the 1920s and 1930s, she felt that her homosexual cousin was an honest fellow and so she started a fight in the Church against prejudices about sexual behaviour. She was also active in a women's church committee to help unmarried mothers. When once asked what she would have done with her life had she been born fifty years later, she was quite clear: 'I would have studied history or social sciences and perhaps I would have become a politician.'

When I came to tell her that I was going to get married and would leave with my husband for Hong Kong, Willemijn (then aged 87) said casually, 'I hope you know about modern ways of postponing having children. It would be a pity when you have a chance to see so much of the world.'

The moment of deepest sorrow in Willemijn's life was the sudden death of her only son, who fell from a rock in the Swiss mountains during a family holiday. After this tragic event, Willem slowly slipped into dementia and died in 1964. Willemijntje followed him a few years later, at the age of 94, reading her newspaper until the very end.

TRUUS HOEKSMA, 1904–

'If God had consulted me only once to know whether it was convenient for me . . . '

My mother, Truus, was born in 1904, the third daughter of Willem and Willemijn. Truus is now over 90; after sixty-six years of marriage she became a widow, when her husband Gerard died in 1994, at the age of 93.

Unlike her mother, and in spite of her strong character, Truus is not a protagonist: 'I would prefer spring cleaning a hundred times over to speaking in public.' The image she creates is one of an extremely sweet, humorous, intelligent, beautiful and very feminine woman who demands little for herself.

By the time Truus had finished primary school, her father Willem and his brother-in-law had created a Christian secondary school. Truus was the first girl to enter and the first girl to pass the final exam in 1922. Truus had wanted to continue studying by taking up History or Language and Literature, but her parents decided differently. Truus complied with their wishes and went to a school of home economics to prepare herself for marriage. She became a cookery teacher.

At the age of 19 she met a serious young man of 23, who was studying Physics at the University of Leiden. They fell in love when he took her out for a romantic trip in a rowing boat. Fortunately he belonged to 'the right faith' (the Protestant Reformed Church), and three years later they were married. Gerard's youth had not been easy. His face was very badly burnt when he was four years old and he had lost his parents at the age of 18. As there was little money available, he went to work for the Philips company. It did not take long for the company to discover that this young man was very bright indeed; they sent him to study Physics on the condition that he would return to Philips after his doctorate. This he did, but at the age of 29 he was appointed Professor in Physics at the Protestant Free University of Amsterdam, from which post he retired only fifty years later. Her husband had an impressive career, and through him Truus met many of the famous physicians of the twentieth century: Einstein, Bohr, Ehrenfelt, Marie and Eve Curie were all familiar names to her. Her humorous observations about people's characteristics instilled in her children a sense of the relative greatness of people of scientific fame.

In 1929 Truus and Gerard moved into a big house in Amsterdam with their first child. From then on, another child arrived every two years: five boys and five girls in all. This was not exceptional among Gerard's colleagues at the Protestant Free University: many of the other professors' families followed the same pattern. All these children were predestined to go to the same Protestant school and of course to the same Free University. Choosing another kind of higher education was a way of branding oneself a rebel.

An important occasion each year was the annual dinner of the Senate: while the professors participated in a formal dinner with serious discussions, their wives organized a party for themselves (also *paid* for by themselves!) in the adjacent room. According to Truus, there was never a sound of laughter from next door, but 'the girls' enjoyed themselves enormously over an (unusual) glass of wine, exchanging the latest news about pregnancies, children and absent-minded husbands.

In spite of having such a large family to manage – her husband present mainly at dinnertime and on Sundays – her children remember seeing Truus angry only once. In the wartime winter of 1944 (known as 'the hunger winter'), the children used to go the houses of the rich to collect waste food. Their father would then cycle out to the farms of his students to exchange the waste food, which could be used on the farms, for food for his children. Returning home from such a trip on a cold winter's day, he brought bacon! He made a fire with almost the last coal in the house. But, absent-minded as he was, something happened and the bacon was completely burnt. Truus exploded into such a rage that the children were far too frightened by her anger to be upset about the burnt bacon.

Truus's compensation to herself, for the large and the small worries of organizing her household, was reading good novels. Her favourites were life stories of 'great' women as well as the famous thrillers of Maigret. Whenever she had a chance, she accompanied her husband to Paris. While he discussed with NATO generals what scientific research was to be done for the defence of western Europe, she wandered through Paris looking for the cafés, the backstreets and the hotels that Maigret visited in search of the solution of a murder. It is no wonder that, on their return home from these Paris trips, Truus's reports were listened to much more closely than those of her husband!

Truus also followed her mother's example of reading the newspaper carefully, especially the birth and death announcements. Later in life, when her days became quieter, she used to philosophize on life and on children, while her hands produced the most elegant and beautiful embroidery.

The most painful event of her life was the death of her second son at the age of 3. She withdrew into the bedroom and did not come out for months on end. It was her beloved brother who finally succeeded in breaking through her self-imposed isolation; he simply went in one day and said to her, 'Come on, Truus, we are going out, I want to buy you a hat.' A year after the death of little Gerard, a new baby came. The family doctor (who assisted at the delivery of nine of the ten children) had told her husband that this was the best way to relieve Truus's grief. It was not. She never got over the loss.

I once asked her how she had felt about all these pregnancies. 'Well,' she said in her characteristically indirect way, 'every time a baby was born, your father and the doctor looked at it and every time they said: "It is a great wonder. Every child is a blessing of God." Of course, I agreed with that. But I couldn't help thinking: if only God had consulted me just once to know whether it was also convenient for me.'

Truus must have been afraid that her children would feel that they were just one of the mass. After the death of their father in 1994, the children had to clear up his library. There was one cupboard which was locked and no key could be found. They broke the lock and discovered two packages in old brown wrapping paper. One of them was sealed with strong, old-fashioned adhesive tape, the other with string. The children hesitated, but finally decided to open the packages. The first one contained the pyjamas of little Gerard who had died, his silver mug, his little bear, some drawings, the death notice and a diary which Truus had written for him. The second package contained a whole set of diaries: Truus had written one for each of her children since 1929, without them ever being aware of it. Here they found precious information on her way of looking at her children, their relationships, all sorts of events, small and big. Her remarks about her children's characters are so lucid that they seem almost prophetic. In the diaries she often remarks that 'the boys' can support each other and 'the girls' likewise. It never seems to have crossed her mind that 'the girls' might be able to offer vital support to 'the boys', and vice versa, which is in fact the case today. Truus philosophizes a lot in the diaries, but there is next to nothing about herself and the way she felt as a woman in her time. Is this the result of her closed character? Or was it a subject about which you simply did not express yourself? Or both?

Truus suffered from hearing problems, which had an important impact on her life; her hearing deteriorated with each of her pregnancies and eventually developed into total deafness. In spite of her hearing-aid, it made intimate contact with her children almost impossible. Not being a dominant person anyway, she would not insist that people spoke more clearly. Rather she would withdraw into herself and observe. In this way she developed a remarkable capacity for perceiving non-verbal signs. Her understanding of people was astonishing, deep and accurate, as those of her children who lived abroad for some time were delighted to discover through her letters.

Afraid to bother others with her own problems, she always tried to find the positive side of things. When I asked her how she managed to accept her deafness, she said, 'Well, it is useful in a way. When the children make too much noise, I simply turn off my hearing-aid and start thinking about the book I am reading. My deafness is a kind of privacy.'

By the time Truus was 85, her memory started to fail and she gradually began to lose coherency in the way she expressed her thoughts. At the beginning of the process certain feelings which she had never clearly expressed about her deafness and her life (suppressed frustrations?) suddenly emerged. ('When you are blind, people pity you; when you are deaf you irritate them.') In this period she also became rather obsessed by a wish to be financially independent. In 1927, when they married, her husband had opened a post office account in her name. Over the following sixty-six years, an amount was added every month. This money had always been entirely her own; but now, at the age of 85, she became adamant about the idea of getting a job. Whenever one of her daughters bought her a dress, she would say, 'No, take it with you. I do not want it. I will buy one when my first salary comes in.' When asked what kind of job she wanted, she would hesitantly reply, 'Perhaps I could work as a charwoman?' Some years earlier, when the first female Minister of Home Affairs had arranged for the State old-age pension to be paid separately to husbands and wives, Truus had sent her a letter of thanks that finally someone had recognized that she too had worked all her life.

Were these signs of lifelong frustrations? Was something sparked off in her mind by feminists insisting that women had to get *paid* work in order to become emancipated? Or rather, was it a manifestation of the culture of the West where the value of work is expressed in economic terms and people's self-esteem is related to the economic value of their labour?

I think I can permit myself to say that my father's love for my mother was the deepest feeling he ever knew. She was in some ways his *raison d'être*. The last five years of his life were no doubt the saddest ones, as she slipped out of his reality to an inner world of her own where he could not reach her. She was deeply afraid of 'becoming mad', and had uncontrolled ways of expressing mistrust and aggression, so unlike her but characteristic of the early stages of dementia. My father could not accept the change in his wife. I once said to him, 'But father, if only you would not contradict her and stop explaining things she can't take in any more, if you would just go along with her in her confused world, maybe it would be less difficult for you.' But he replied angrily, 'No, that I will never do, because then I would not respect her dignity any more.'

He suffered deeply. She suffered deeply. They could not share their suffering any more. They moved to an old people's home, too late to put down roots again. But they felt each other's presence. They clung on to it, holding hands, day and night. And then, one morning – for the first time in his life – Gerard refused to get up. He stopped eating, he stopped drinking, and finally he did not want to be touched any more. Truus saw her husband quietly slipping out of her reality. The evening after the funeral she said to me, 'It was a beautiful ceremony today. What a pity Gerard could not be there.' In the first weeks after his death she seemed to know what had happened without clearly understanding. But she felt the loss of his hand in hers. Now she is lost.

She does not recognize us as her children any more, although she senses that there is something familiar about us. Her most joyful moments are when her small great-grandchildren come to visit her. They don't have a problem with great-granny. She is just sweet and laughs in her own humorous way.

A few weeks after her husband's death, she asked me, 'Are you married?' – 'Yes, mam' – 'Do you know him?' – 'Yes, mam.' Silence. 'Do you know what lonely is?' – 'Not really, mam.' Long silence. 'Lonely . . . loneliness . . . is . . . that you can't give anybody a name.'

The last time she had a flash of full consciousness about her condition in my presence was when I was going through a family photo album with her, something she usually liked to do. Suddenly she said, 'Put it away. I do not know these people anyway.' – 'But, mama, you do know who you are yourself, don't you? Look at this photo, this is you, mama, Truus.' She looked at me intensely, bewildered, her eyes expressing deep anxiety. Slowly she said, 'No, no, I do not know, I do not know who I am, I do not know who I am . . . ' Then she started to cry, desperately, something I could not remember ever having seen her do.

A lifelong devotion to husband, children, grandchildren and great-grandchildren. And when you really need them, when they visit you more often than ever before, you do not know what the relationship is. You do not know who you are. In bygone days, she used to recall lines of poetry that had struck her. One which she recited more than once was a poem of Rilke which ended

> . . . for in the deepest things of life,
> one is alone.

EDITH SIZOO, 1939–

Overcoming the fear of the otherness of others by trying to meet them

Home

I was born in January 1939. The fact that I was the seventh of my parents' ten children was largely responsible for my feeling that, as a person, I was nothing special. I did not make trouble, so I got very little personal attention from my parents; my eldest sister played an important role for me as a kind of substitute. For a long time I had difficulties determining my own worth. The first important event in this respect was the birth of my youngest brother. After six weeks he was moved out of my parents' bedroom, as was the custom. There was a rearrangement of the children's bedrooms, and this little baby came to share the smallest bedroom in the house with me. As an 8-year-old, I was to take care of him in the night. The significant moment came as a result of a particular habit of my father's: every Saturday he would buy 100 grammes of very expensive chocolates, just for my mother. One Saturday he brought the same chocolates for me. This gesture greatly enhanced not only my self-confidence but also my feeling that I liked being a woman.

Aside from this event and, later on, the family's appreciation for my cooking, I very much remained just one of the many. I saw at friends' houses how different it could be; but I felt very sorry for those who had only two or three sisters and brothers, and took care that they came to dine at my house. To me a table of twelve was the norm for enjoying a good meal. Home was a haven – cosy, safe and often lots of fun.

The fact that I was endowed with good brains was not considered anything special, either. You were simply expected to do well in school, and I do not remember my parents ever asking whether we had finished our homework. It was my mother who, in the background, insisted that the girls would get just as much education as the boys – not because we performed well at school, but 'because you never know what may happen in life'. My father, however, was quite resistant to one of my sisters who wanted to become a medical doctor ('too rough a profession for women'); nor did he agree with my wish to go to the theatre academy ('the artists' style of life is not ours'). My sister did not give in, and became a medical specialist. I did, and went to university to study French Language and Literature. Although my parents were proud of the professional performances of their daughters, they remained ambivalent about them combining a family and a job, not to mention my going by myself on four-week field-trips to other continents, leaving the children with my husband. At a period when I was working full time and earning almost half of my nuclear family's income, my father said, 'Of course, I understand you like to have some spare-time income, but your first responsibility is the children.'

Second World War

Another determining influence on my life was the impact of the Second World War. The war began affecting The Netherlands when I was 18 months old: my very early childhood was increasingly consciously coloured by fears, insecurity, soldiers, war stories and never having enough to eat. The atrocities against the Jews, which I witnessed as a very young child, and the growing awareness in later years about what had really happened during the Holocaust, instilled in me a very acute and profound feeling of indignation about oppression by racism. As a young woman I was much more preoccupied with that than with inequality between men and women.

One day, when I was about 5 years old, I witnessed a *razzia*. The Nazis dragged Jewish families – people of our neighbourhood – out of their houses; long silent rows of frightened Jewish people passed by our house. A pregnant woman carrying a young child in her arms fell. A soldier kicked her hard in the belly, snatched the child from her arms and flung it against a tree, again and again until there was no life left in it. These and other events left me with an unanswered cry: Why them, why not me? I couldn't understand. I did not see a difference between the Jews, myself and the Germans.

It is the Holocaust that sowed the seeds of my never-ending search for ways and means of overcoming the fear of the 'otherness' of others – a fear that, it seems to me, lies at the root of racism and perhaps at the root of every form of will to dominate others.

Religion and ideologies

It was also the experiences of the Second World War that triggered off my profound suspicion of any form of institutionalization of prophetic messages, political or other ideologies. In my teens and twenties I felt infuriated by injustice and violence, but I was not a militant activist. I was seduced by the ideas of the Left, but I was suspicious of the way in which my leftist friends started to think in terms of systems, universal strategies and models, and the power play going on in those circles simply disgusted me. I refused to conform to the rules imposed by a group in order to be accepted. I was deeply afraid of new uniforms. I never wore blue jeans.

At university (1957–1962), I belonged to the generation of trailblazers of the turbulent 1960s. It was the period of decolonization, the war in Algeria, the new 'mission' of 'development'. We discussed the Protestant faith and the new evangelism of Marx. But very few of us were as yet ready to break with the Church and its imposed ideas about what we should believe and how we should behave.

At that time I was not yet fully conscious of what really disturbed me in the Protestant expression of the Christian faith: the separation between the rational and the mystic, the lack of joy and beauty. The churches were sober and functional; the sermons were geared towards interpreting Bible texts. The fact that each of the many Protestant churches claimed to have a better interpretation than the others, while the Catholic Church claimed to have the one and only true faith, roused in me more doubt than confidence in what the content of that faith should be. I rejected completely the idea that people of other religions would not 'go to heaven'. In the institutions that had 'organized' the faith, I saw power play, fights over translated words, and an emphasis on Eve and her sisters as temptresses, to be kept away from the holy tasks which would be performed by men only. Whatever arguments male theologians had invented to justify this seemed to me (literally) man-made.

I did not find what I was looking for: an expression of the deeper mystery of Life, the kind of spirituality that transcends the limitations of the rational and the cerebral; something that makes you experience being part of the mystery of the Creation; something that expresses my deepest feeling that the Life within me is stronger than the body I am living in. It was only through my experiences of Oriental spirituality that I began to discover exactly what I was missing. To me the Protestant tradition of that time appeared to be painfully void of feminine ways of living the mystery of Life.

Enlarging my world

The best decision I ever took in my life was to marry the man I am still in love with. I do not know any other man I appreciate more for just being the way he is, full of subtleness, humour, discretion in personal relations, with his way of asking very little space for himself while creating a maximum for others to be themselves and to grow. I have not met any other person who has stimulated me more to accept myself as I am, and to flourish.

Immediately after university, we left for Hong Kong where he had to set up a food programme for refugee children who had come by the millions from mainland China. We came into contact with all layers of society in this British colony: Chinese refugees who ranged from university professors and big business men to illiterate street cleaners. We learned their language, which proved to be a most fruitful way of establishing contact with them and learning about their culture.

Although Hong Kong was quite an experience, the real cultural shock came in India. I cannot think of any other cultural surrounding that is better able to call into question all that seems universally evident to you as a European. India: its contradictions and multitudes of truths, beautiful philosophies and the caste system, the chastity of women and the sensuality of the sculptures in the temples, the normative 'discourse' and the injustice of poverty. India fascinated me. It was also India that awakened me to the issue of the oppression of women. I liked and admired the women in India and I became increasingly amazed by the way in which a large part of society – men and women alike – seemed to have internalized a denigrated conception of women. Through my eyes, their strength, resilience and beauty contradicted this conception in its very essence.

The most important influence which India had on me was the realization that my ways of life, my norms and values, are far from universal. It was there that I experienced the beginnings of an understanding that people may perceive their realities differently, that their 'strategies' to improve their situations are determined not only by their socio-economic environment, but at least as much by the complexity of their visions of life and death, their beliefs, symbols and values.

This awareness continued to mature when, after our return to The Netherlands, I was confronted with debates on 'development' in Europe and the 'strategies', 'models', policies and criteria for the financing of projects that ensued from it. There appeared to be a yawning chasm between the ideas of all those eloquent 'developers' in the rich Northern countries and what I had experienced and observed while living among 'the poor' in Asia. Poor in what? Rich in what? Certainly not 'rich' in the capacity for listening and for changing one's own perspective on the development of others.

Professional life

After four years at the Ministry of Foreign Affairs, where to me the atmosphere could be summed up as 'much ado about nothing', I switched to the NGO world and became General Secretary of a Netherlands' Federation of (thirty) NGOs for development cooperation. I was the only female director in the NGO sphere in my country, but I have experienced this fact more as an asset than a constraint. In so far as there were men who showed their unwillingness to accept a woman in a leading position, I felt that this was their problem not mine. There were three things, however, that I found (and still find) very tiresome:

- The fact that many men (particularly those in leading positions) are more preoccupied with power than anything else.
- The kind of 'double-thinking' I have to do all the time: I usually was (and still am) one of the very few women in meetings. Consequently the discussions and the underlying power play are enacted within a masculine framework of operation. I find myself reacting to what is said with a silent 'Yes, but . . . ': 'yes' I understand what is said, 'but' I have to integrate it first into my way of perceiving reality, interpret it from that frame of reference, translate my understanding and feeling about it back to their mode of expression (lest I am accused of being irrational, intuitive, etc.), and by the time I come forward with an opinion the men have already moved to the next point. So I am known for showing delayed reactions.
- In the job just mentioned, the second in command was a man, who could not accept emotionally that he worked 'under' a woman. My present male colleague and I are on a perfectly equal footing. He accepts that and we have an excellent working relationship, recognizing each other's specificities. In both cases, however, the outside world almost automatically presumes that the man in the office is the director and I the secretary. I doubt that men have had the experience of being regularly reminded, through gestures, telephone calls and letters, that it is taken for granted that in an office setting the man is the subordinate and the woman is (part of) the leadership.

I left my job with the Netherlands' Federation of NGOs after twelve years, because I felt that in spite of my influential position I had not succeeded in bridging the gap between my experiences in Asia – which were profoundly cultural – and the Western way of thinking about development and cooperation between the South and the North.

I left a secure job, a position with social status, to go to Brussels and work for the South North Network Cultures and Development – no financial security, no status, but working with people from all parts of the world on a different basis and touching on fundamental issues in South–North relations; searching for ways and means to overcome the fear of the 'otherness' of others by trying to facilitate the building of bridges between people. Professionally, I am finally at home.

My nuclear family

Our first child was born in 1964 in Hong Kong with the help of a Chinese nun. She had a wonderful way of massaging the waves of pain out of my body. I remember her repeating all the time, 'Be strong, you are going to have a son!' And finally there she was: Solange Myriam. We were delighted with her. My husband, announcing the good news at his office, was told, 'Bad luck, Sir, better luck next time; you are young enough!' And so we had a second child, this time in India: 'Bad luck, Saab, better luck next time.' Manon Shanti, 'peace of mind'. And we were delighted again. The third and fourth times we were 'lucky': Robert and Jeroen. The family was perfectly balanced.

My children have taught me an awful lot about being a woman. They made me discover the invaluable privilege we have over men: to receive the Life for our children to nurture and to feel it grow to maturity as an integral part of ourselves. I think it is the most existential experience (in a positive sense) which a human being can have. Once born, they also appealed constantly to the feminine qualities in me – and not only in their early childhood. They are young adults now, ranging from 24 to 31 years of age.

In spite of the fact that their mother has a university degree and has had some responsibility outside the home for the last twenty years, they tend to address mainly their father when their studies, politics and matters of professional life need to be discussed. Sometimes the implicit denial of what I might have to contribute, beyond support in times of trouble in personal relations, makes me wonder. I try to see it as their way of balancing their parents' specificities in meeting their needs. In any case I find myself feeling excitedly happy when they come home for a relaxed chat and when I see how these four beautiful, self-confident children relate to each other, to us and to the world they are living in.

SOLANGE, 1964– AND MANON, 1966–

'I thought women's issues were a problem of my mother's generation until I started working'

'Women of my generation are ambivalent as to what they want to be'

Solange was born in 1964 in Hong Kong, and Manon in 1966 in India. In 1969 and 1971 their brothers Robert and Jeroen followed. Having been moved around by their parents, they finally settled in The Hague, a relatively quiet, green, bourgeois city characterized by the presence of the Dutch government and its administration.

An important event in Solange's life was returning from India to The Netherlands. Her 'culture shock' was the discovery that, suddenly, the language she had spoken exclusively with her parents, their secret code, was being used by every person she met. Almost instantly Solange closed herself off to outsiders and it took her many years to open up to them again.

All four children went through the Montessori education system from the ages of 4 to 18. Both Solange and Manon feel that the education they received at home and at school was based on a love of (and respect for) who they are, and was geared towards their growing into maturity as people:

> Distinction between boys and girls was so absent that we were convinced that women's issues were problems of the past, at least in our part of the world. Thanks to your generation we have exactly the same rights and opportunities as boys and we take them.

After secondary school Solange went to university to study Art History. Most of her fellow students in this faculty were girls. She has had the same boyfriend for the last five years. She has become fond of Amsterdam: friends, cafés, canals, bicycles, bookshops, theatres and little money; relaxed, dynamic, progressive. After graduation, she joined a small publishing house which produces photographic art books. The firm is run by three directors: Sybrand (a man), Paula (a woman) and Solange.

> It was only then that I started to discover that men simply have no other choice than to accept the presence of women in the workplace. But that has not changed their mentality. Even the so-called progressive ones manifest deeply rooted traditional conceptions of women. They may listen to your opinion, but they expect a man to be in charge. The fact that outsiders always assume that Sybrand is the director and Paula and I the secretaries is only a minor symptom. It's just the way they approach us. When we make it clear that the three of us have the same kind of responsibilities . . . sexist jokes galore . . . When Paula and I are around, Sybrand usually explains the situation, but . . . well . . . he rather *likes* the misunderstanding . . . if you know what I mean. It is a syndrome. I feel that women have to be very strict with men in this. Constantly. I do not trust any one of them in this respect. The problem is that for us there are other things in life which we find just as important as a career. We are not really prepared to invest 200 per cent of ourselves just in a career. And many men do. So in the leading positions, they will continue to be in the majority and masculine values will continue to dominate. They like power more than we do. We also want children, but there is a lot of insecurity: job insecurity, relationship insecurity, and the question of what kind of home this polluted earth can provide for future generations. Personally, I want to achieve something professionally first and assure myself a position. Then I can permit myself to have a baby. Perhaps.

After secondary school, Manon went to the Academy of Arts in Rotterdam. After graduation she obtained a special scholarship for young artists in Amsterdam. Now she has set up a collective artist studio in Brussels with two artist friends. Her boyfriend, also an artist, is living in The Netherlands. Although she loves Amsterdam, she feels that you need to expose yourself to other surroundings, too. Manon emphasizes that for her generation 'identity' is associated much more closely with the individual person than with a social group, political affinity, church, extended family, etc. While you do not feel responsible for a group, you do feel responsible for friends or the family members you like.

> This individualism has its positive and negative sides. In partner rela-
> tions you do not stay together because you are obliged to, but because
> you want to. You are more interesting for each other when you both
> develop your own talents. You feed each other in that respect. The
> negative side is that the younger generation seems to find it very diffi-
> cult to live closely together, especially in small places (most of us have
> very little money).
> It is not society that restricts possibilities for women. It is women them-
> selves. In the world of artists, for instance, it does not make any difference
> whether you are a man or a woman: it is your work that counts. But male
> artists on the whole are much more market-oriented and better selling
> themselves than female artists. What I also see happening all the time in
> love relationships between artists is that the woman very easily makes her
> own career interests subservient to those of her partner. In order to create
> things, you need isolation. And women artists give in to their tendency to
> care. For women there is more need to make choices than for men because
> of their biological nature. But women of today are ambivalent. They make
> half-choices and then they accuse the men of exploiting them. In fact they
> are unhappy that their choice does not allow them to flourish fully in both
> ways. You have to make clear choices and then find a form of living in
> accordance with that choice. In my social surroundings it is almost taboo
> to say: I choose to live exclusively for my husband and children because
> that is what makes me happy. Still I feel that that, too, should be accepted
> as long as it is a conscious choice. Women's lib? Among my girlfriends we
> do not see the need for women's movements any more. None of us is a
> member of a women's organization. But we are very keen on having close
> girlfriends and we cherish these friendships. Perhaps . . . we need them
> even more nowadays than in our mothers' time.

REFLECTIONS

Personal and external influences

What can be concluded from the ways in which the women described in this contribution were, and are, shaping their realities, is necessarily a reduction of a complex pattern of motivations. It seems to me that the first and predominant factor is a strictly personal one: character. With regard to external factors one could say that the Protestant Christian environment influenced the lives of Willemijn, Truus and Edith considerably, although they did not react to it in the same way. For Willemijn, a rather rebellious character and a socially committed woman, the Church offered – in addition to the faith – an opportunity to move beyond the borders of her home. For Truus, who was not a social activist by character, the Protestant Church environment served more as a social and religious frame of reference. The validity of these norms and values were not a matter of serious doubt for either of the two. The same framework, however, pushed Edith to look beyond its borders for spiritual nourishment from other sources. For Solange and Manon, religion is simply not one of their preoccupations.

Another influential external factor may be the changing views towards women carrying out paid tasks in society. Willemijn might have found more fulfilment in her life had she been given the opportunity of professional training and paid work. Her daughter Truus, however, rather than expressing a desire to participate in shaping society outside the home, always gave the impression of being quite fulfilled with her large family. Her obsession with finding a paid job when she was 85 seemed to be related to her sense of self-esteem. She often referred to the fact that her labours as a carer were never officially acknowledged as worthwhile (hence her joy at receiving her own State pension). In this sense, fulfilment and self-esteem may not necessarily be interdependent. Edith has shaped her life in such a way that she finds fulfilment as well as self-esteem inside and outside the nuclear family, although she experiences the combination of the two as a physical and psychological constraint. Solange and Manon feel that it might be wiser to select one's priorities in life more clearly and stick to them.

The sweet and sour fruits of women's lib: choices of today and tomorrow

In my part of the world (Europe, The Netherlands), the post-Second World War (and especially post-1960s) wave of emancipation has not only had the positive result of 'liberating' women from a restrictive role in society sanctioned by secular or ecclesiastical laws and male fears. By contributing to the sexual revolution of the 1960s, and by succeeding in obtaining equal rights for women and *inclusion* into sectors of society which used to be mainly or exclusively reserved for men, women's lib has, paradoxically, imposed a new constraint on the women of today and tomorrow.

This time the constraint seems to be more complicated than those imposed on the *secluded* woman. In addition to the continued pressure on women to be seductive, understanding and caring, they now feel that they also have to prove their ability to function in positions which were traditionally reserved for men and which were shaped according to male values. It is increasingly difficult for women to confine themselves to a 'woman's world' of activities and values, so that the expected and accepted behaviour of women is becoming less and less clear. The key question has become: How can one retain a feeling of personal integrity as a woman who functions in worlds of differing values?

Women of today and tomorrow are faced with new concrete choices in the private as well as in the professional sphere – choices which call into question the values of yesterday. In the private sphere these choices pertain to three kinds of relationships:

- *Adult relationships* Heterosexual/bisexual/lesbian? Short-term/long-term? Living-apart-together or living together? Formal marriage/partner-contract/non-formal? Monogamous or accepted occasional extramarital relationships?
- *Parent–children relationships* To have children or not? With/without a father? Crèche or home care? Abortion when child is unwanted or defective? And, in the future: to predetermine the sex of a child or not?
- *Child–parent relationships* Over the last forty years there has been a tendency to hand over responsibility for 'the weak members of society' (the old, the sick, the handicapped, drop-outs) to the government, by way of the State social security system. Nowadays we are witnessing a shift in this tendency: with the economic recession and neo-liberalism gaining ground, State social security systems are being called into question for financial reasons. In concrete terms this means that the informal social networks of family, friends, communities, etc. will have to take over (part of) the task of caring. And so other choices will have to be made: Do I take my old parents into my home? Who should take care of them: myself alone, or my partner and I together?

When entering the professional world, women are confronted with a pattern of organization and rules of the game determined by underlying values which happen to be mainly masculine. The choice then is: Do I adopt these, adapt to them, compromise, or reject them? How do I deal with formalized power when I am not in a position of power, or (more difficult) when I have acquired a position of power? How do I get my priorities right (family, career, participation in the community, cultural interests)?

Women in my country are living in a strongly secularized society of which the organizational design is increasingly shaped by economic neo-liberal criteria and less and less based on values of solidarity and social justice. One would be inclined to think that, in this context, the forces that determine the choices that women are making in their search for security, are those that guarantee to the largest possible extent socio-economic independence.

Although we see many signs of this, it would be jumping too quickly to conclusions to leave it at that. As the neo-liberal economic world vision excludes increasing numbers of people from the paid labour force, this situation might, in the long run, become normal rather than abnormal. Meaningful ways of shaping this reality will also become more and more vital. Work ethics will change; so will social security ethics. People will become much more dependent, once again, on 'connections'. Their need to belong to some kind of group will become stronger when there is no workplace to 'belong' to. In this sense women's age-long experience in devising strategies for fulfilment and self-esteem by creating informal social networks which provide security and significance in their lives may prove to be of great value in the twenty-first century.

In the dual society of today and tomorrow, women's role is twofold: to feminize the workplace where paid work is carried out, and to use their experience to shape the reality of non-paid but vital contributions to society.

16
GIVING AND TAKING SPACE

Jaana Airaksinen, Finland

Maija, 1900–1989

Anne, 1939–

Jaana, 1961–

MAIJA, 1900–1989

Searching for wisdom, allowing change

Rosa Maria (Maija) was born in 1900 in Rauma, an old port town on the west coast of Finland. Her father owned a small stonecutting business. Her mother was from a big farming family. Maija had an older sister and a younger brother. She always considered her father a nice and gentle man; almost too good, and her mother she considered tough, with a sharp tongue and penetrating eye. The family lost all their money when the father's business partner disappeared with it: Maija's mother started washing clothes for others, Maija herself worked in a grocery store and her sister emigrated to New York in search of work. Her father died in 1918, at the end of the Finnish civil war. Maija paid her own way through school, and she and her mother supported her brother in getting an education and a proper profession. Maija also bought him a violin so that he could play music: music was something she enjoyed throughout her life.

Maija was engaged to a man from Rauma. They loved each other and had plans for a future together once he had completed his university studies in Helsinki. One day her fiancé came home and told Maija that he had had a relationship with another woman, who was now expecting his child. He told Maija that she was the only one he loved, and asked her still to consider marrying him. But Maija refused, even though she was deeply in love with him; she felt that the other woman would become an outcast, and her life would be difficult and burdensome if she was left alone with a child at that time.

Maija was determined to re-establish the family's position in society which, they felt, had been lost because of her father's excessive kindness and benevolence. She had decided very early on to have her own profession, which would bring her

208

respect and provide her with her own income, but would also enable her to enjoy her life intellectually. With her dreams of love and marriage shattered, she became even more determined to go ahead with her plan to become a teacher. She again paid her way through the training, and got a job as a headteacher near Rauma. Her mother, meanwhile, had sold the house and given the proceeds to her son for him to start a business. Since she needed a place to stay, she moved to Maija's house.

Maija cherished the memory of her fiancé until she died, although they never met again and she refused to marry anybody else. Although this was a time of romantic love, her choice was rather unconventional. Most women still achieved their place in society and their financial security through marriage; it was also the only accepted setting in which to have children. It was out of the question for Maija to have any sexual involvement outside marriage or any close relationships with men if not aimed at marriage. As a teacher she was very much a focus of attention; she felt obliged to comply with the behaviour expected of her and to set a proper example to her pupils. As she grew old and realized that her earthly life was almost over, she felt that she had really missed something and deeply regretted it. She gave this much thought, and used to advise us (her granddaughters) that if we 'did not get a partner from heaven' we should 'accept one from the earth'.

Life went on. Maija became well established and busy. She was the centre of her community's social life, always organizing church choirs and charity campaigns, children's plays and theatre visits. In 1945, at the end of the war, there came a surprise which was to change her life completely: a 6-year-old girl marched into Maija's house and told her that she was looking for a home. After some days Maija agreed that Anne could stay, despite her own mother's initial disapproval at the idea of having an extra mouth to feed. Teachers' salaries were meagre, but they were given housing at the school and a small plot on which to grow vegetables. Maija started to sell her produce at the market and to keep a few animals for extra money, and to supplement their income further Maija and her sister ran a boarding-house at the sister's farmhouse during the summer breaks. Every summer they would pack their belongings and move to the countryside for three months.

Maija's finances were tight until late in life, but she worked hard to expand her property. For her, accumulating wealth was one's obligation towards the subsequent generations, and an essential part of being able to feel secure. Needless to say, she worked hard throughout her life doing the housework, running the school, growing her own food and keeping the boarding-house. Her hands were always busy mending, baking, cleaning, making handicrafts, comforting and caressing. When she sewed a dress it was never just a plain dress, but always had delicate embroidery. She was creative and resourceful in collecting beautiful pieces of work, which she cherished: some things she made herself, some she saved for, and many were given to her as gifts.

Maija retired from teaching after Anne was married in 1960. She was delighted about the marriage, and the status of Anne's husband, a doctor, gave her additional pleasure. Maija retained a deep respect for him until the day she died. Maija and her sister started to spend the winters with Anne's family in Savo province in eastern Finland, and after the birth of Anne's two daughters they moved to Savo. They had their own apartments, next door to each other and near to Anne's house. Maija took daily care of her grandchildren – we played, did handicrafts and sang songs. She quickly adapted to her new surroundings, becoming involved in church and music activities, and had a circle of friends around her.

Maija established strong social networks wherever she lived, and her interaction with people was rather extensive. Her visitors' books are full of beautiful stories and joyful memories of people getting together. Her hospitality was well-known, and she often had friends and relatives staying at her house for long periods. These visits were reciprocated, she and Anne staying with relatives in other towns, enjoying theatre trips, concerts and other social events. She also kept regular contact with her former pupils and boarding-house guests, and had a wide and interesting circle of friends, from sea captains and missionaries to writers and theatre people. Through this circle of friends, she accumulated a collection of exotic items such as Tasmanian porcelain, baskets from Ambokavango, innumerable books and long letters.

The Lutheran Church and religion were a solace and source of meaning to her. Throughout her life Maija was a regular churchgoer and an active member of her parish. She sang in church choirs, organized sewing circles and other fund-raising events for missionaries. She practised such Christian beliefs as putting others before oneself and loving oneself through making others happy. Her Lutheran ethics placed high value on work and social justice, and her beliefs were full of mercy, based on the thinking that those who render shall receive. Serving others seemed to be a genuine pleasure to Maija: it was rare to see any resentment in her. During her last years, the fear of having been selfish and of the judgement that would imply surfaced. She was afraid of dying and spent many years trying to find an inner peace.

Her sister died in 1974 and Maija suffered a small stroke after the funeral. Life suddenly showed tangible, concrete limits and she felt a terrible sense of loss, adding further to her anxiety. She spent a lot of time contemplating her relationship with her sister, feeling keenly her own inadequacy – regrets were constant. In the end, no longer able to manage, she moved into Anne's house. By that time, Anne and her husband were both working long hours and we children were in our teens. It was not easy for Maija to adjust to our hurried lifestyle, and she must have felt lonely and swept aside at our house. We all lived together until her physical condition no longer allowed her to be left alone – she moved into an old people's home in 1987.

The move was very difficult for her: at first she thought she was in a hotel with Anne and kept calling her. It was a time of distress for Maija: she started to live more and more in the past and wanted to move back to Rauma, where she thought she would feel at home again. Anne visited her regularly; those of us who lived further away saw her only occasionally. She died in a hospital at the age of 89.

Maija had strong beliefs and values and, until late in life, a healthy self-confidence. She was persistent and would rarely give up: when she was told that Anne could not be adopted, she went up to the President and his wife to get the permission, spending days waiting to see people who could assist her. And she succeeded. She had a positive outlook on life and a sense of humour; she made the most of situations and enjoyed people just as they were. She knew her own surroundings well – the place and the people, the traditions and the culture – and she appreciated them. She physically belonged. Her life was full of excitement and activity, but balanced by a flowing rhythm and silence. Until she was old she would wake up early every summer morning and walk from the main house, through a birch alley, down to the white sandy beach; there she would take a long swim, alone, in tranquillity and peace.

Religion was important to her, as well as the church; they gave her security and understanding, but in old age she became terribly worried about the final judgement. God became the grand judge whose mercy and love were hard to gain. She died without being sure whether she would be forgiven, although just before dying she showed signs of being more at peace with herself and her life.

Maija participated in women's grassroots organizations such as church groups, choirs and sewing circles. These were the first forms of women organizing themselves in Finland. She took responsibility and accepted leadership. The fact that she was not married gave her exceptional freedom to make decisions about her own life. In many ways, she chose to lead a very different and more independent life than her own mother and than many women of her age.

ANNE, 1939–

Looking for protection, resisting change

Anne was born in 1939 in Rahikkala in the Soviet Union. She had one elder brother and two older sisters. Her mother died when she was 2 years old. She belongs to the Ingria people, a Finnish-speaking minority which was located in the Ingria land around St Petersburg. It was a German-occupied area at that time (during the war), and its inhabitants were looked upon with great suspicion and were purged by the Soviet regime. Their house lay on the outskirts of the occupied territory, right next to the Soviet-held St Petersburg, and was subject to nightly harassment. Their father, fearing arrest, kept the family ready to flee at any moment – one evening in 1943 he received a warning, and they fled after nightfall. The son, who was out and could not be reached, was left behind.

After three months' journey on foot, through the woods, they reached the

211

Estonian coast and took a boat to Sweden. There they were directed to Helsinki and put into a refugee camp. One of the sisters, who had curly blonde hair and big blue eyes, and whose head had not been shaven like the others with black hair, was quickly placed in a family. The father got a job in a metal factory in Rauma, and the oldest sister started working, too. Anne was placed on a farm where the mother mistreated her: she became very sick and was hospitalized. After she had recovered, she stayed for a while with the doctor, who helped her to find a new family. She was 6 years old when she met Maija, who gave her a home and adopted her a few years later.

The members of Anne's family were able both to stay close to each other on the west coast and to maintain regular contact. The Russians – the term then used to refer to all Soviets – were often openly disliked, so the family did not make much attempt to preserve their Eastern tradition, nor to keep up with Russian, which was their second language. The brother had settled in Estonia and managed to establish contact with his family in Finland in the late 1950s.

After the war, some Ingrians returned, some stayed. However, times were uncertain and this influenced Anne deeply: she was constantly afraid of a forced return to the Soviet Union, until she got her Finnish citizenship at the age of 15. Those fears haunted her for a long time, and still return in certain situations (such as border checks). It was not until the late 1980s, long after the rest of her family had made the trip, that she felt able to visit her brother in Estonia; even then she was fearful and jumpy when crossing the border. Situations involving pressure were always difficult for her to handle; although she was aware of Maija's expectations for her to go on to higher education, she could not bear the idea of going through university exams.

She did, however, travel to Germany and contemplated studying there. But neither Maija nor Anne could bear such a separation and she returned to Finland. She began a practical cooking and home economics training, found that she enjoyed it, and started thinking about a career as a Home Economics teacher. Meanwhile, she met Pekka, her husband-to-be. Pekka was finishing his medical training and was keen to get married quickly, which they did. His first assignment was in Lapland, in northern Finland: Anne gave up the idea of becoming a teacher in order to follow him. She stayed at home and spent her time knitting dozens of pullovers. But she found the long and dark Lappish winter such a sinister experience that she soon announced that she would move, whether he joined her or not. Pekka agreed and they moved southwards to Savo province (where he came from and where most of his family continued to live).

Anne still had some aspirations for a profession, but her husband was strongly opposed to them living apart. In their small village there were no possibilities for studies and she gave up the idea again. Their first daughter, Jaana, was born in 1961 and the second two years later. When she was very young, Anne's oldest sister had given birth to a son outside marriage; this nephew moved to Anne's house at the age of 15. Anne always had help with the housework and was never fully occupied with it. With Maija, she had been brought up in an environment where everybody worked, where any job was considered better than idleness, so she soon went back to work outside the house.

The inner insecurity stemming from Anne's childhood, a problem which she had never really dealt with, led her to build her own life in ways which made her feel emotionally and physically very secure. She created a fortress, inside of which she would feel safe with the rest of her family, keeping the unknown 'them' at a safe distance, or denying that they existed. Maija was a source of harmony and security for Anne; she gave unquestioning love and acceptance and sheltered her from events in the outside world. Anne felt that she owed Maija everything, and this deep sense of gratitude caused her many feelings of guilt when she realized that she had to balance her obligations to Maija with those to her husband and herself, that she could not reciprocate Maija's unconditional devotion. Anne transferred some of her dependency from her mother to her husband. She feels that he is an integral part of her, that one without the other is not complete. She accepts and follows her husband's opinions about public issues such as politics, even though (when she really starts to think) her true feelings can be quite different from his. However, she seems not to mind this contradiction.

Anne is invigorated by her social networks and by the family. The feeling that she belongs, that people depend on her and that she can depend on others, is of fundamental importance. Social relations and emotions are considered her domain. She thinks that everyone in the family has a unique role and explicit characteristics which make us fit that particular role. These characteristics are stable; we remain more or less what we were twenty years ago. The strength of this is that everybody is allowed to have certain good and positive qualities, which the family enjoys and depends upon, but certain bad and destructive characteristics are recognized as well. The positive qualities get a lot of support and nurturing, the negative ones are openly acknowledged and talked about. Children and grandchildren are part of her feeling of strength: her daughters achieved something professionally to which she aspired, and the grandchildren are fresh springs of joy in the continuum of her life.

She shows her caring by giving gifts according to her own liking. When she goes to buy a coat, she will immediately buy four: one for herself, one each for her daughters and one for her daughter-in-law. She watches carefully to see whether or not the gifts are appreciated. She demands appreciation. Her worth needs to be stated by others through their explicit gratitude. Just as she was utterly grateful to her mother, she expects her own children to show gratitude towards her. She sees herself in comparison with others, almost in a competitive setting, and suggests that she is better – more generous, more beautiful – and those closest to her have to reassure her that this is the case. She shares her life and allows you no choice: you have to share it with her, in the way she wants.

She talks a lot about her emotions and shows them freely within the family, but she cannot tolerate confrontation from the family members. Her children's puberty – when, in our attempt to become independent, we openly and aggressively challenged her values and the way of life she had built up and cherished so much – was a time of great distress still vividly remembered. Nevertheless, she has managed to keep the family together. She wholeheartedly supports and defends the members of her family – when her sister bore her first child outside marriage, Anne talked about it openly and convinced everyone (including her own father) that this was all right. Her world is either/or, good or bad, us or them, a black-and-white place where 'they' are always a potential threat. There is little space to breathe and no room for otherness, except to be rejected.

She has created a very safe environment for herself. Money has been an important factor in bringing her security and fulfilment; it enables her to have some control over her life. She has accumulated wealth in her name. She goes prepared – if something unexpected happens, her whole life will not be in shambles. Beauty and harmony are important to her. She looks for beauty in people's appearance and in the way things look. Her own foreign looks have drawn positive attention and have been a source of self-confidence. Looking older is difficult for her, and she invests time and money in trying to preserve her face and body. She cleans all ugliness from her environment: the house is spotless and tidy, she closes her eyes to images of war and sickness.

She was happy when her children left home and her daughters went to university, but when Maija moved into the house in 1982 Anne entered a difficult period. Maija was getting old, the impending separation invoked anxiety and sorrow in Anne, and she felt deeply obliged to Maija. She had to balance this with the stress that Maija's continuous presence put on her own marriage. Her husband, used to their own privacy, did not find it easy to accept the old woman's intrusion into his sphere. Letting her mother eventually go into an old people's home in 1987 was a traumatic event for Anne, and a source of guilt for years.

Maija died in 1989 and (with all the children out of the house) Anne began to have more time for herself. She took on additional charity work and spent more time with her husband at the old farmhouse they now had as their summer home. We used to spend all our summers in the country, at our cottage beside a lake. Somehow that cottage was in the middle of Nature, and yet removed from it. We never walked in the many surrounding woods and we never really appreciated the earth and what it bore. We just indulged in distant admiration of the scenery. In recent years Anne has been spending much of her time enlarging and taking care of the garden, deriving considerable pleasure from all the different flowers and birds around.

Lately Anne has given up the heavy burden of worrying and has relinquished some responsibility for her children's lives to the children themselves. These days, she is more balanced and at peace with herself. Feeling more secure as a result of her age and her experiences, she is better able to respect herself, and thus also able to show her respect for others and demonstrate a little more tolerance for differences. Her husband retired recently, but she decided to continue working for a few more years, until retirement age.

Anne never questioned the idea of combining family and paid work. She did not consider them as conflicting with each other and did not feel guilty about sharing the daily care of her children. She firmly believed that she was entitled to be happy in her own right and that only then could she make the people around her happy, too. Although she always had an air of strength, she was both strong and weak at the same time. She did exercise great influence on our family life – taking complete care of the finances, for example. Anne brought up her children for work and for independent life: we had to study and we were expected not to settle down too early. She raised her daughters with a clear idea that we should have our own profession and be able to support ourselves financially. At this point in her life, she is happy and content. She says that the only decision she would change, were she able to do it all over again, would be to get a proper education.

JAANA, 1961–

Trying to understand: life is change

My childhood was happy and stable, and very secure. We spent a lot of time with the thirty or so cousins from my father's side who lived nearby, and often had relatives from a distance staying at our house. There were always people around. Within that, our family was tightly knit, with well-defined roles for everyone to guide our behaviour. My mother was the emotional one, and strong. When she was young she could smash plates against the wall in anger, or storm out of the house slamming the door and swearing she would never come back. But we all knew that within two minutes she would turn round and walk back in, hugging us all while tears ran down her face. In a peculiar way, she was proud of this behaviour, so unlikely in those surroundings. She reserved for herself the right to be expressive, as she put it. With her side of the family, there was always a lot of noise and talking, crying and laughter.

My father was strict and serious and emotionally distant, but we considered him very fair, rather like a respected judge at his elevated desk. We always thought twice before approaching him. There is, however, one particular image of my father being susceptible which I fondly recall: the family returning late from a party, me sitting on my father's lap in the car, half asleep; he relaxed and joking lightheartedly, and smelling of whisky and cigars. It was these passing moments of softness that I clung to. All his side of the family were raised to have qualities of endurance and perseverance. They were hard-working, especially the women, and they all had strong opinions, were stubborn even – the kind of people who do not speak to one another for thirty years after a dispute. Only the women, with their small solid statures and hardened hands, seemed to have some joy and laughter to lighten their burdened lives. The men were a solemn bunch and there was a heavy air around their strength. My father never got rid of that, even though he had decided that the toughness which he and his many siblings were brought up with was not going to be repeated. He taught us to develop our rationality and logic and always to trust science beyond anything. He passed on to us his strong personal ethics and moral integrity; social justice was something he practised and was part of his life. He believed that everybody ultimately stands alone and that one must therefore rely only on oneself when striving for something – and naturally you should constantly be striving for more. With mother we had a sense of belonging to a succession of generations; we were part of a tradition, albeit a willingly changing one. In father there was little appreciation for his roots and few stories to go with his large family.

It was my father's responsibility to take care of me, while mother looked after my little sister. I was often told that I was just like my mother, while my sister's personality is similar to that of our father. I think it was safer to think that way. When mother was working, our grandmother, Maija, took care of us. She was soft and warm and full of stories, a truly accepting person. I remember her encouragement, the sense that so much was possible. Of course, we took her for granted and it was only later that we learned to appreciate the time we spent with her. Looking back, I feel that she was a very balanced person: she used her intellect and her heart and was always working with her hands, too. She was the one who taught us how to use our hands, how to knit, do embroidery and tend flowers.

My parents were obedient and respectful citizens, avoiding change. They emphasized justice and honesty and there was no hypocrisy whatsoever in them. These values were never questioned or discussed. They were simply there, heavy and solid and very black-and-white. It is a long unlearning process for me to understand that life is a much more complex affair and to see these opposites as part of one interacting whole, with nuances and degrees – to be rational, in its original meaning. For different reasons, both our parents kept us very sheltered from the outside world. But inside, the different traditions were interacting, opening up possibilities for surprise and allowing for constant contradiction, which I realized later was a source of creativity, too.

When we moved to the nearby provincial capital in the middle of a school year, the only class which had space for me was the one meant for students with learning difficulties. I had always been top of my class, dutifully learning my homework by heart, well turned out and innocent with a tight ponytail. Suddenly I was in the midst of people with problem-ridden backgrounds. I was fascinated, and spent all my time with my classmates learning about a whole different world. But I also felt deceived, in a way, at not having been told that life is so complex and varied and sometimes so cruel. I slowly began identifying with, and became part of, what had been so vehemently rejected and denied by my family. I became the Other, yet I could not be just that. My parents felt lost and anxious.

I started spending the summers in Sweden, working and living in an immigrant neighbourhood. Prejudices ran through my head: I never knew I was so full of them. It was puzzling. I took off to the USA on a scholarship and got acquainted with a Latin American group. I vividly remember our guitar nights with patriotic love songs and old rebel tunes.

I did not have any clear plans for the future, but somehow thought this was the point to decide. That it had to be university studies was self-evident: I chose Economics. This was happily supported by my parents, as we girls were always encouraged to go into traditionally male professions, such as medicine, law or business, and discouraged from female ones. But I found very little stimulation at the school I had chosen; I was ethically out of place there and directed my energies elsewhere.

This coincided with a several-year process of separation from my live-in partner, from whom I also found no stimulation and with whom I was psychologically at odds. It was a stressful time; we became petty and cruel to each other, resorting to all kinds of sordid games just to get the other one to leave. I remember feeling so sore and so numb at the same time, feeling a bottomless well inside me providing a constant outpouring of tears. I recall hoping that once my eyes started floating they would carry my body and mind away, too. My straw at that time was a job opportunity in Papua New Guinea, and I clutched at it willingly.

We were still unable to let go of the last remnants of our relationship and, against all odds, I became pregnant. I still think that this only shows the decisive influence of our subconscious in determining the course of our lives. Recalling how happy and strong I felt about the child, I think that deep down I must have been waiting for just that. With all this energy inside, I was finally able to break out from the situation. We were, nevertheless, now bound to each other through our baby. At the start it was a loaded bond, carrying all sorts of colliding expectations of parenthood and traces of being hurt for such a long time ourselves. Ugly scenes took place and I was surprised – after the relief of having freed myself from an incessant fight for space to do and say what I thought was right – to find myself in a battle over my body. My partner somehow must have seen my body (and my inseparable soul?) as just some kind of delivery mechanism for his baby, and therefore to be in his possession. He took the right to intrude as his privilege. His self-righteous manhood was outrageous. We were back in our game of getting at each other, only now there were three of us on the scene. This was, however, a decisive change, as my attention was focused on trying to protect the innocent one thrown into the midst of the battle. Respect – whatever way it may be expressed, so essential in any relationship – crumbled away and vanished. When there is no respect I do not believe there can be a relationship beyond a forced pretence.

My daughter was born under lucky stars on a Sunday morning and had pointed ears, like all fairies. And we had ever more reasons to disagree as the weight of the differences in our values became magnified by the sharing of parenthood. Everything was a matter of principle, and therefore of great importance. Child support – which I still think *is* a matter of principle and, if nothing more, of symbolic meaning – we ended up settling in court.

I married a long-time friend the year Kukka was born and we all travelled together to my next posting in Malawi. We had a wonderful time there and our

217

daughter was born two years later. My greatest relief was when my spouse adopted Kukka and I no longer had to worry about her fate should something happen to me.

We both became frustrated with our work and felt that we were slowly being swallowed up in a machinery alien to, and removed from, the everyday lives of people in that location. It was like being in a minibus full of people driving over cornfields, crushing well-tended land as we transported 'development'. We left and moved back to Finland, where I enjoy the feeling of physical and cultural belonging once again. I have a history and a voice here. It gives me profound joy indeed.

It has taken me a long time to begin to listen to myself and to be able to rely on my strengths and weaknesses as part of a whole, to respect myself and to be able to respect others. It is only recently that I feel I have created enough space for myself to breathe and to contemplate, and I am now better able to let others around me breathe as well. As I am no longer afraid of losing myself in the process, I have sensations of real connection with others. In a way I feel truly autonomous: knowing that you can be separate and that you are not dependent, you can allow people so close to you that the boundary becomes blurred or momentarily disappears. My children have contributed greatly towards this understanding.

Despite the fact that relationships with one's children are unique in the sense that one rarely contemplates whether or not one will remain in the relationship – it is, in a way, a quite unquestioning love, an enduring relation. Living so close together requires a constant and conscious balancing act. This relationship carrying so much responsibility within it provides a sense of the limits to our actions; the fulfilment too is riddled with feelings of inadequacy and therefore exposes limitations in ourselves that we would otherwise not be aware of. It is because of this moderation of our actions and the feeling of responsibility connected with them, this expansion of knowledge of ourselves and the dreaming of dreams beyond our own happiness, that we can be rewarded with most precious and thorough understanding. Children do bind us to life through all kinds of small practicalities and enable us to see the many miracles nearby.

I am slowly growing out of the restlessness which never let me stop. I am beginning to let go and to understand that without sorrow happiness just isn't complete. Learning to appreciate differences, and to see contradictions as a source of creativity and therefore life – in its deepest sense, to allow space and silence enough to hear one's own rhythm – is a long journey. I feel balanced with my life and somewhat detached from outside expectations; I am becoming me all over again.

Part III
THE FINDINGS

17

HOW WOMEN CHANGE PLACES AND PLACES CHANGE WOMEN

Edith Sizoo

INTRODUCTION

To try to present an analysis of the life narratives which comprise Part II of this book is both challenging and full of pitfalls. The material is life material. It is like an ocean filled by rivers from different points of the compass, moving forwards and backwards with currents and undercurrents which merge into each other, difficult to grasp. Every effort to take hold of the currents, to pin down the waves, runs the risk of forcing the material into a pattern which reduces its fullness. No attempt will be made to provide an exhaustive discussion of the multifarious themes running through the narratives. The following will focus on only a few major issues: some of these were distilled from the life stories themselves, others emerged from discussions between the authors during their Encounter which carried them beyond the narratives.

TIME AND PLACE IN WOMEN'S LIVES

Place as a particular mixture of social interactions

The notion of 'place' has been perceived and defined in many different ways. In the social sciences the more *static* notion of place equated with 'community' (which was supposed to be bound by a shared set of values and common rules for behaviour) has now been replaced by much more *dynamic* conceptions. It may not be coincidence that this has occurred at a time when spatial distances have been drastically reduced and time barriers broken. Relatively new terms like 'the global village', 'globalization', 'time–space compression', 'a global sense of place', or new meanings for old words like 'the Market', indicate a feeling of being connected to a world far beyond the boundaries of 'a place called home', even if this may be experienced in a volatile way.

Doreen Massey (1994) provides a clear and useful analysis of these changes, which lead her to conceptualize places as connected by social relations stretched through space as well as having their own local characteristics:

> If one moves in from a satellite towards the globe, holding all those networks of social relations and movements and communications in one's

head, then each 'place' can be seen as a particular, unique, point of their intersection. It is indeed a *meeting* place. Instead then of thinking of places as areas with boundaries around them, they can be imagined as articulated moments in networks of social relations and understandings, but where a large proportion of those relations, experiences and understandings are constructed on a far larger scale than what we happen to define for that moment as the place itself, whether that be a street, or a region or even a continent. And this in turn allows a sense of place which is extroverted, which includes a consciousness of its links with the wider world, which integrates in a positive way the global and the local.

(Massey, 1994: 154–155)

With this conception of 'place' in mind, it is challenging to look at the life narratives from a diachronic, multigenerational point of view (from grandmother to granddaughter within a given geographical context) as well as from a synchronic, intragenerational angle (the same generation in a particular period of time, but in different areas of the world).

The meaning of 'place' over time

The first question, then, is how the mixture of local and wider social interactions particular to place influenced the lives of the grandmothers in the narratives as compared to their daughter's and granddaughter's lives.

It seems that over the generations the influence of the wider social relations has changed more in terms of *quantity* and *nature* than *intensity*. The *quantity* of contacts with wider networks of social relations has obviously increased over the years. The *nature* of the wider social relations has mainly changed with regard to worldwide ideologies – from the unavoidable influence of institutionalized (world) religions and a personal experience of 'faith' (as a place-transcending relationship) towards increased influence of more secularized (worldwide) sets of ideas (be they Marxist, feminist, ecological or other). However, it remains to be seen whether the *intensity* of the latter is as pervasive as that of the former. It is hard to determine yet what the bearing of worldwide secular movements will be on the youngest generation's feelings about the meaning of 'place'.

In terms of a *sense of belonging,* the assumption that in the grandmothers' time local relations were more determining than social interactions with the wider world is in fact only partly confirmed in our collection of life narratives. The women of *all* generations put almost equal emphasis on the impact on their lives of family relations, friends and various kinds of 'communities' (village, school, associations, movements, etc.). On the one hand, from generation to generation the women in the stories experience fewer – physical as well as mental – barriers for moving beyond boundaries: the granddaughters are less compelled to negotiate social space to fulfil their aspirations than their grandmothers were. On the other hand, the granddaughters' need for belonging seems to be more consciously experienced. The way they express their sense of belonging differs from their grandmothers, in that it is perhaps less identified with a place-bound community. However, this does not automatically imply that the youngest generation feels 'place-less'. Rather, they satisfy their need for belonging by consciously searching for and building up a variety of social networks of different spatial dimensions and contents.

An illustration of the struggle for 'belonging' appears in the story of Yvonne Deutsch, a Jewish Israeli woman, who left her grandparents' extended family house in Romania. That house to which she never returned still means 'home' to her. At the age of 8 she went with her mother (who had been abandoned by her father) to Israel, a foreign land, where belonging and peace meant identifying with a militaristic culture of survival.

- In her thirties Yvonne became directly involved in the Women's Movement in Israel which opposes the military culture. It was the *intifada* that brought about active opposition against the Israeli occupation of Palestine territories. Every Friday, Women in Black stood at different busy public sites in the big cities of Israel and engaged in silent protest by holding up a cardboard hand saying 'End the occupation'. The colour black stood for mourning. It was felt that the Jewish people belonging to the Zionist State of Israel had failed to synthesize their own experience of loss, destruction and extermination. The Women in Black consider this failure dangerous because it prevents the soul from healing the fears and traumas of the past. They see the use of military force as a destructive form of compensation. This act of female protest gave rise to strong negative reactions from within the Israeli society, including from Israeli women. But it intensified some women's awareness of the need to establish a female political culture of peace based on women's life experiences as an alternative to militaristic values including war, oppression, exploitation, violence and rape. This political, as well as feminine, claim also brought the Women in Black into direct contact with some Palestinian women who – contrary to many Israelis – understood the message. Yvonne now feels that she lives in a cultural vacuum: as an adult she rejected the political culture of Zionist Israel and now wants to be integrated into the culture of the Middle East. She is seeking a Middle Eastern women's culture. She wants to belong, somehow.

The experience of places at given periods of time

By bringing together life narratives from a variety of geographical areas, it is possible to see the significance of a particular period of time, in which the same generation in different parts of the world experiences 'place'.

The life narratives provide ample illustrations of the impact of particular periods of time on the lives of women. The generation of the grandmothers, who were young around the beginning of this century, shows striking differences in the various geographical areas. The grandmothers in Southeast Asia, the Middle East, Africa and Latin America (those of relatively well-off families, as well as poorer ones) were more exposed to the wider levels of social interaction, and in that sense were less isolated than grandmothers of comparable backgrounds in Europe. This may be due to the fact that colonialism had considerably more impact on colonized societies than on the European ones in terms of widening social interaction between different cultural settings of the globe: at that time 'Europe' was much more present in colonized countries than vice versa. The families living in Pakistan, India, Sierra Leone, The Gambia, Sudan, the Philippines, Ecuador and Brazil could not avoid being confronted with Western behaviour, education, religion and commercial relations. Their influence is expressed in all the stories from those countries. In the European life narratives of the elderly generations, however, there is no mention of lifestyle and beliefs being put into question by Africans, Hindus, Muslims or Indigenes. This may also explain why the stories of the European women are somewhat simpler than those from the colonized regions of the world, which have been affected by colonial and post-colonial politics and values.

Wartime conditions like those in Israel/Palestine have different effects on women living in that area. In Palestine the struggle for independence has opened opportunities for women to move beyond the boundaries of the home and affirm their capacities of fulfilling functions in society. In Israel the same condition of perpetual tension and threat rather affirmed the women's nurturing role. They were 'programmed' to internalize the oppression of the Palestinians as justified and necessary for survival.

CHANGING PLACES

Moving between places

Experiences of women moving between places are abundant in the life narratives. Whether freely chosen, of necessity or imposed, for long or short periods, painful or enriching, such experiences are always of great significance. Leaving home means packing one's past experiences into a suitcase, unpacking them in another spot under the sun and discovering that they light up in a new way.

• Edith (The Netherlands) tells how she felt dissatisfied when attending Protestant church services as a child. At that time she did not know exactly what made her feel uncomfortable. It was only after she had moved to India, where she experienced Oriental spirituality, that she began to discover what exactly she was missing.

• Gowri, who grew up in India, had gone through higher education and had travelled abroad extensively before she went in for an arranged marriage in her mid-twenties. She had carefully reflected on the choice between a 'love match' and a marriage arranged for her by her parents. She says that, having spent long periods in the countries of the West, she had decided that for her the popular Indian system was better after all.

Mobility is often associated with providing opportunities for women to widen their horizons and achieve greater autonomy. Although in many of the life narratives this effect is clearly demonstrated, they also show that there are prices to be paid.

• Durre's (Pakistan) grandmother was born in Sialkot, which was in the (then) undivided British colony of India. She was literate in her native language. At the age of 14 she was married to a divorced man with two children. She had to move to his parents' house in another state of India. Her powerful and very conservative mother-in-law treated her as little better than a slave, while her husband wanted her to be 'modern'. Being caught between the two, she lost. Her husband fell in love with a 'modern woman' and she had to go back home, where she served the household for the rest of her life.

Durre's mother, though not keen on marrying, given her mother's experience, agreed to marry a medical doctor twelve years older than herself, mainly because he was planning to go to Europe in order to specialize further. She says that 'In a sense, it was a matter-of-fact decision based essentially on the assumption that I was exchanging one cage for another (but perhaps bigger) one'. In Europe he taught her many 'worldly things', from eating with a knife and fork to an appreciation of Titian and Beethoven. At the same time, he expected her to conform with highly orthodox, traditional, forms of behaviour. She felt bitter towards her in-laws, who would have preferred the marriage not to have taken place, but her husband made it clear that if she were to get along with him she also had to get along with them.

When Durre herself was 17, her parents decided that she should get married to a cousin. She initially resisted the idea, but finally gave in, mainly for the reason that he was going to Europe. The marriage did not work.

The price to be paid for physical mobility is also apparent in the mental dilemmas which professional women often face when separated from home by their participation in the paid labour force. This can create tensions in terms of consistency between the inner world and the outer world. This is what Edith called the 'sweet and sour fruits of women's lib':

When entering the professional world, women are confronted with a pattern of organization and rules of the game determined by underlying values which happen to be mainly masculine. The choice then is: Do I adopt these, adapt to them, compromise, or reject them? How do I deal with formalized power when I am not in a position of power, or (more difficult) when I have acquired a position of power? ... The key question has become: How can one retain a feeling of personal integrity as a woman who functions in worlds of differing values?

• Itisal was born in 1973 in Palestine. After her parents divorced, she first lived with her mother and later with her father: '[Her] generation no longer has to deal with the problems of dress, education or, to a certain extent, with sexual relations at work. But the pressure of society still prevents them from taking decisions about their personal lives . . . [She studied] Business and Administration, she now works in Jerusalem and is developing her skills further through various courses . . . She has many male penpals from other countries, and has taken part (as one of very few women) in meetings between youngsters from Palestine and Israel in England . . . She has come to the conclusion that she cannot trust men, and feels disgusted by the idea that they look at her as a body only . . . Sometimes she does not know what she wants and needs, which makes her angry. This influences her relationship with males; building up a love relationship is difficult.'

The stories of the Indigenous women from Ecuador and Brazil clearly demonstrate that when mobility is imposed, holding on to a sense of rootedness may help to retain a feeling of dignity.

• The ancestral land of Eliane Pontiguara's (Brazil) nation has been taken away from the people by the Portuguese invaders. Her people are perpetually on the move, out of economic necessity. After her father's death, she moved with her mother to her grandmother's house in the poor immigrants' zone of Rio de Janeiro. 'My grandmother created an Indigenous ghetto inside the house and we lived in a cultural cell . . . She wanted me to study . . . [but] forbade me to talk to the other children [in school] . . . One day my grandmother gave me a green transparent stone. That green stone was the link between generations in my Indigenous family; with it, I received all the culture, all the tradition and all the spirituality that these women carried all their lives. I used to travel in that stone. That was the positive side of my maternal house-prison – it was in fact my Indigenous village, in exile from its own land.

• At the age of 11, Christina (Ecuador) had to leave her village in the mountains to go to school in the city. After a while, she escaped from the convent school run by German Catholic sisters because 'It was a prison, simply horrible. I felt as if I did not exist any more. There were four other [Indigenous] girls there as boarders: we were treated by the sisters as objects, not as persons, and certainly not as personalities . . . It was strictly forbidden to talk Quechua.' Slowly she became aware that what the Catholic sisters were doing to her now had been done for ages to her people. And so she decided to climb over the wall of the convent and join the Indigenous movement. Some years later she married a Belgian development worker and went to live in Europe. There she dedicates her life outside home to the defence of the rights of her people and raising awareness among Europeans of how Western ideas about 'development' are destroying the dignity of her people in Ecuador.

Staying in a changing place

As has been observed by Janice Monk and Cindi Katz (1993), 'the reverse of mobility is not always immobility'. Women who spend most of their lives in the same location may perceive that place quite differently in various periods of their lives. What makes the difference here is not them moving between places, but the outside world moving into their place. For some, these influences may widen their own mental horizons. For others, they may create mental or physical isolation: while more and more members of the family or the community are extending the boundaries of their world physically or mentally (education, travelling, the Internet, etc.), they themselves are left behind with a feeling of backwardness, uselessness, loneliness. Although in our collection of life narratives one finds more examples of the first effect, the second one should not be disregarded or underestimated.

> • Ethel Crowley (Ireland) mentions this sense of being left behind and losing touch when she describes how emigration 'has been the main link between Ireland and the rest of the English-speaking world throughout the twentieth century'. Her mother's family was typical of the 1950s, which was a period of deep economic recession in Ireland: four of her five brothers left. None returned. She stayed, as it was more unusual for single young women to emigrate alone. And emigration is still part of 'the national psyche': of the six brothers and sisters with whom Ethel grew up, four have left Ireland: 'The situation emerges where siblings have to rely on snippets of information relayed in occasional letters, photos and telephone calls in order to keep in contact. Divisions emerge between those who stay and those who go, between those who associate with an idealized view of Ireland from afar, a land frozen in time, and those who live in it. Parents suffer and, above all, mothers suffer. They sometimes regret having borne children in a country that is underpopulated by European standards and yet somehow cannot provide a livelihood for its daughters and sons. New ideas about globalization of the economy and of popular culture tend to devalue home . . . The pain of separation is obscured.'

THE INTERTWINING ACT

A destereotyping effect

In the course of the polylogue, the presentation and analysis of the life narratives had a gradual destereotyping effect. For instance, the popular distinctions frequently attached to 'North' and 'South' (such as: rich/poor, assertive/subdued, modern/traditional, active/passive, etc.) appeared to make little sense with regard to women's perceptions of themselves and of their role in society. The descriptions of the women in these fifteen stories, in deprivileged as well as privileged situations, in countries from the South and North, serve to unravel stereotyped images of women. Although they do not manifest grandiose heroic self-images, neither do these women define themselves as typically passive, weak, oppressed, irrational and sacrificers, as opposed to active, strong, powerful, rational and dominant. And there are, of course, a great range of characters, from the subtly strategic to the confrontational, from the inwardly pensive to the outspoken.

Even under conditions of physical abuse and fear for the life and existence of their communities, whether in the southern or northern hemisphere, hardly any of the women in the narratives describes herself as a victim. Without denying the presence of oppressive power structures, what these women saw as important were the forces that gave their life a meaning and a sense. There was great diversity in what these stimuli and sources of energy might be. It became evident that one's 'identity' is contextual and, in that sense, multiple. Women, just like men, change their priorities according to situations and moments of time.

The narratives reveal a blend of strength and vulnerability. The women in the stories often combine a feeling of not being fulfilled or stimulated to their full potential with a determination to fight for (and often achieve) what makes life worthwhile for themselves and for others. They often relate imposed 'responsibilities' with a consciousness of their 'ability to respond'. Their role of caring for others is a question of choice rather than compliance. However, many do have to contend with the fact that the task of caring for others is imposed upon them as if self-evident – something which many do not appreciate.

Some of the individuals in the stories choose to function in the public sphere: women do not always shy away from leading positions that imply power. However, they are not willing to take on such roles at the cost of other things that contribute to their personal fulfilment. In contrast to the image created by some internationally known women who have occupied powerful political positions, for many of the women in the stories the problem with 'power' seems to lie in keeping the balance between 'power over' and 'power to', in an arena where the former is more frequently sought after.

Whereas some Southern feminists have reacted to a Western feminist discourse representing '*the* Third World woman' as a category of human being that passively accepts oppression and victimization, the life narratives suggest that feminists worldwide could be blamed for comparable generalizations. By stressing so publicly and so insistently the subordination of women, feminists in both North and South have implicitly reinforced the distorted image of women as passive accepters of this situation. Their often indiscriminate use of words like 'empowerment' has contributed to that image, implicitly suggesting that women lack 'power', thus veiling the fact that 'power' is a multifaceted phenomenon which may take many different forms and is exercised in many different ways – including

the possibility of women exercising power over other women in an oppressive relationship. One example that challenged stereotyped images of the relation between
women and power came from Safia Safwat's Egyptian–Sudanese family:

• Safia Safwat (Sudan) was born of an Egyptian mother and a Sudanese
father. Her grandmother on her father's side was the daughter of a very
wealthy slave-trader. In spite of the fact that slavery was officially abolished
in the Sudan in 1933, her grandmother's slaves stayed with her until she
died, as they had nowhere else to go. The grandmother was a powerful
woman who, as the head of her large extended family and as a rich patron
of men and women alike, was used to being in charge.

She had longed for a traditional bride for her eldest son as this would
have enabled her to arrange a traditional marriage which involved elaborate, exotic preparations and endless feasting and exchange of gifts. Her
son, however, wanted to marry a foreigner, an Egyptian girl. She never
forgave him, and took it out on her daughter-in-law: it was partly her
constant complaints that ensured the ultimate failure of the marriage.

Safia's mother married Safia's father at the age of 15. In spite of resistance from her own mother and her mother-in-law, Safia's mother had a
strong belief in education for girls and taught her sisters-in-law to read and
write. After a painful divorce she chose to take up the education of her
four children herself. In spite of the extreme wealth of her in-laws, Safia's
mother drew on her own social, emotional and financial resources to bring
up her children. She stayed in the Sudan and took up dressmaking. She
managed to get her children through secondary school and university.

Safia did a Ph.D. in Law. She joined the Attorney General's Chambers
as a legal counsel and was chosen to represent those chambers at the prestigious national Law Commission, where she was the only woman. Her
political activities earned her the wrath of the regime and she had to leave
the Sudan for London. There she lives with her husband and two sons.
She practises as a lawyer and teaches Islamic Law and Human Rights at
the University of London, hoping that one day she will return to her
homeland and be in a position to apply what she is teaching now.

Moving beyond dichotomies

Although the narratives posed challenges to assumed patterns of similarities and
differences among women, such similarities and differences still remain. Apart
from an emphasis on family life, one of the most striking parallels between the
women in the narratives was the importance they attributed to education, for
both girls and boys, and to the personal urge – or even fight – for education for
themselves.

The deeper and more intriguing similarities, however, were found in the underlying currents, in the forces which drove the women of the stories to shape their

realities the way they did. Differences between them became manifest in the forms in which these driving forces were expressed in concrete, everyday reality.

While participants in the polylogue discussed their difficulties in dealing with dominant ways of conceiving reality, they discovered basic similarities in their own reactions and those of the women in the other stories.

It was felt strongly that, in dominant ways of defining reality, phenomena which are in fact composed of different, interrelated elements are too readily dichotomized. This tendency leads to distorting pictures of reality as divided into, for instance, rationality vs. intuition, scientific knowledge vs. popular wisdom, the social and spiritual vs. the physical, private vs. public, male vs. female, development vs. underdevelopment, and so on. These distorting dichotomies are often associated with a priori value judgements, one side of the dichotomy frequently being seen as superior to the other. Furthermore, dichotomies are all too often presented as consisting of opposing forces, where one must give way to the other. The creation of these opposite categories is an act of separation, which results in the separated categories acquiring a status in themselves.

In fact, however, these categories are pure constructs and do not reflect reality. Opposites exist only in relation to each other and are apparent in differing degrees depending on the situation. They are interacting parts of the same whole, as symbolized by the yin/yang principle. They correlate and their boundaries are constantly moving.

An analysis of the narratives seems to suggest that many women in the stories (without explicitly saying so) do not necessarily agree with this tendency to value the one category more than the other, nor with the inclination to oppose the two in a potentially destructive manner. In concrete life situations, some women go even further, tending to relate the two 'opposites' with each other in such a way that they become not only complementary but also mutually enriching. By so doing, they move beyond the dichotomization of reality.

The reluctance, or even refusal, to simplify reality by reducing it to mentally and practically more manageable proportions may also explain why many of the women described in the life narratives are combining several purposes and drawing on different strategies at the same time. This way of acting on their perception of reality may be seen as a sign of '*confusion*', but can also be interpreted as '*fusion*' of the many facets of life. The following examples may illustrate this observation.

Interweaving attachment and detachment

• Kamala Ganesh, born and living in India, describes the way in which her husband's grandmother (1906–1969) discovered her special gifts to assist people who were in search of physical relief and spiritual growth. She became a guru and started her own ashram. Contrary to the practice of male gurus, the members of her own family also lived there. She fused her roles as mother (who creates and nurtures a home) and guru (for whom the entire world is a home), dissolving the boundaries between home and world and yet maintaining a distinction, interweaving detachment with attachment. She was able to reach out to the public and also plumb deep into her own inner self.

The stories about women in situations of severe material poverty and exploitation, or psychological oppression, show intriguing similarities of response. In spite of physical suffering and the threatened annihilation of one's feeling of being a person in one's own right, many women in the stories are witness to remarkable efforts to resist this erosion of the self. Relief is often sought in relating the personal identity to a shared group identity; in entering into a close relationship with non-human manifestations of life, such as nature; or in symbolic signs of one's dignity as a human being. Striking descriptions of responses to this identity problem were given by Esperanza Abellana from the Philippines and Nicole Note from Belgium.

Keeping the balance between personal and shared identity

• The Filipina women described by Esperanza Abellana (Philippines) are not exceptional: the lives of many poor women are stories of physical and mental hardships. Women taking up the hard task of responsibility for the family end up by not being able to separate their own identity from it. They are so used to denying themselves that they seem to have become a nonentity. Nevertheless, it is not unusual for them to become active members or leaders of community organizations. The woman's feeling of 'wholeness' seems to lie in the extent to which she can keep a balance between her personal and shared identity by putting the personal in a wider perspective, connecting personal needs with collective interests, identifying herself with both and experiencing them as intertwining.

Relating the self to nature

• Maria (Belgium), although born in Flanders, spent her early youth in a village set up for Belgian soldiers stationed in Germany. After this carefree period, she returned to Belgium where the restrictions of an urban environment and a rather rigid Catholic education curtailed her sense of wholeness, joy and freedom. She started to perceive her reality and herself in a negative way and developed an inferiority complex. She reports that her relationship with nature, experienced in deep silence and loneliness, has helped her throughout her life to gain the strength to continue:

Sitting in the grass at noon, with only the sound of the larks, the heavy heat embracing her, the warm smell of the earth surrounding her – only then did she feel her batteries recharging. She had experienced the same feeling of harmony at her godmother's farm. There she would spend

231

hours on her hands and knees, searching for fresh potatoes in the dry earth, the sun shining down on her back. Being at one with nature gave her strength, fulfilment and purpose. It made her complete.

These experiences had, and still have, a determining influence on Maria's perceptions of life, the education of her children and her priorities for society.

Stories told by Eliane Pontiguara (Brazil) and Kamala Ganesh (India) illustrate the way in which women in situations of extreme deprivation respond by first taking care of the immediately useful, while at the same time connecting it to the symbolic with a view to safeguarding their dignity.

Poverty and the use of limited resources as symbols of dignity

• Eliane is a daughter of the Indigenous Nation of Pontiguara, inhabitants of land in what has been called Brazil since the arrival of the colonizers. Now the Indigenous Nation of Pontiguara does not have land, nor a State, and even their language was lost in the course of colonization and modernization. Still, the people of Pontiguara feel that they belong. Eliane describes how their belonging is expressed through the simple dishes they prepare, symbolizing the dignity of their people: Mandioca is a basic food for the Indigenous people of Brazil. It unifies them. It is dignity. Dignity is built step by step. It must be watered every day, just like mandioca.

• Kamala's domestic help lives in the slums of Bombay. She has two young children. During the Hindu–Muslim clashes in Bombay in 1992, her hut was burnt and she lost all her belongings. She came to Kamala's house and told her what had happened. When Kamala asked what she could do for her, what she needed most urgently, she answered: 'An iron'. Why an iron? Was there nothing more urgent than that? No. She wanted to wash and iron the few saris she had left, since otherwise the crumpled sari would expose her legs, thus impinging on her sense of dignity.

These types of responses seem to correspond with the ways in which many women throughout the ages have shaped their realities: using limited resources as symbols of dignity, sharing their personal fulfilment in life with the collective interest of the family, finding sense in caring for the offspring of humankind and sensing the overarching wholeness of nature. Often, these responses may also reflect a survival mechanism: the void left by the erosion of one's individual identity is filled by living through others and/or through different manifestations of life, as in nature. As is sometimes evident in times of major crisis, it may be because of the small, immediate necessities and concerns that people survive. In this sense responses to conditions of oppression, non-recognition, physical and mental marginalization show similarities.

The life narratives of this book also provide ample examples of ways in which women integrate, rather than separate, 'tradition' and 'modernity' in their ways of life.

Integrating tradition into modernity

• Durre Sameen Ahmed was born and raised in a deeply religious Muslim family in Pakistan. Today, she and her two children still live in an extended family set-up of four generations. In her youth, she learned to ride a horse, shoot with a rifle and play the piano. She was unhappy at her convent school, which was run by French nuns, but says that the misery of school was compensated for by an atmosphere at home which placed great emphasis on what was considered the best of Western culture: music, literature, thinkers and a general reverence for science and knowledge. Although her parents also tried to inculcate in her a familiarity with Eastern culture, this was largely overshadowed by a genuine respect for the West. Durre studied in Pakistan, Europe and the USA and became a psychologist. As she approached her forties, she began to realize, mainly through her practice as a psychologist, that many 'modern' ideas of knowledge were in headlong collision with the spiritual side of humans. It was a knowledge which asserted that 'God is dead' and which viewed human nature in extremes of black and white. She began to realize the extent of this dichotomy, as it existed not only in modern psychology but in her own social and intellectual relationships as well. Thanks largely to her mother's extensive knowledge of spirituality as related to Indigenous/Islamic conceptions of health and healing, Durre managed to retain both her faith and the considerable intellectual discipline she had gained in Western universities.

Coming to terms with a desire for both professional career and motherhood appeared to be a matter of concern for most of the women in the stories aged between 18 and 50.

Fulfilment in profession and/or family?

• In Shanti George's (India/The Netherlands) family, embedded in the historical and cultural context of the Syrian Christian community in the Indian State of Kerala, female professionalism (medicine and sciences) has been part of the tradition since her grandmother's youth. Shanti's mother, a medical doctor, chose to combine professional and domestic tasks. She sees 'compromises' between her work as a medical doctor and her family as 'trade-offs' that have enabled her to combine professional as well as domestic fulfilment.

Shanti herself, at the age of 32, was appointed a Reader in Social Anthropology and Sociology. She resigned from this position in order to

marry and move abroad. She has had a series of varied assignments, but
no longer a tenured academic position. Now that she has the new commitment of parenthood, and has decided not to take up employment outside
the home for the first part of her daughter's life, it may be that she will
never have such a position again. Shanti says that she does not see the situation in either/or terms: although she would not agree that her life choices
represent a step backward from some feminist agenda, neither does she try
to present them as a step forward. She prefers to see this period of her life
as a step sideways, enabling her to explore new choices and combinations
in addition to familiar ones, and to move away from notions of 'forward'
and 'backward' on some sort of linear scale where 'forward' often
connotes individual achievement through competitiveness.

• Jaana Airaksinen (Finland) is combining a professional career and a
young family. She sees the concepts of 'private' and 'public' getting new
meanings and being redefined. If we see the world and us in it living in
interactive, mutual relations, many concepts become more versatile and
have more layers to their content. For example, an autonomous person can
be defined as someone who is not afraid of losing herself/himself into
others and is therefore able to loosen the boundaries and melt into other
beings, rather than as a person not making connections.

Moving beyond boundaries, relating personal interests with collective needs,
making choices and trying to keep the balance – these should not be understood
as a way to be constantly in harmony with oneself and one's surroundings, or as
an effort to avoid risks and conflicts at any price. The life narratives provide
many examples of courageous, often painful, acts of breaking away from
expected behaviour. These disconnecting acts, however, often result in recon-
necting in other ways: with oneself in terms of one's self-confidence, priorities,
spirituality, and even with one's supposed enemies.

Gaining self-confidence by steadfastly facing conflict and pain is the thread
running through the stories provided (among others) by Amal Krieshe from
Palestine.

Breaking and relating

• Amal's (Palestine) grandmother, Mas'uda ('lucky woman'), lived in the
north of the West Bank in Palestine. She was married to her cousin at the
age of 16. Her husband died young during the popular uprising against
the English mandate. She herself was illiterate, but sent her sons to the city
to study. When her second son graduated from the University of
Damascus, she went to visit him – something no woman in her village had
ever done. The political activities of another son gave her a further reason
to stand out: it gave her the self-confidence to talk about politics. Her independence generated hatred between her and her daughter's husband as

well as with other women in the village, because she was an unusual woman who had achieved things that others had not.

Her daughter Nabiha ('the clever one') was not allowed to go to school and was married to a close relative at the age of 18. When her husband decided to build a house in the village and charged his wife with the supervision of the construction, she came into conflict with her mother and father. But her self-confidence grew and she discussed public affairs in front of men. When she decided to go on a literacy course, this brought her into serious conflict with her husband and her social circle.

Amal's mother (Nabiha) freed her from wearing the head cover. Grandfather, father and brothers were furious, but she persisted. Amal found much self-realization through school, extracurricular activities and (in particular) through theatre. She scandalized her family by participating secretly in a TV play. When her father refused her request to go to university, she went on hunger strike . . . and won. At university in Jordan, she had her first love affair, which ended when she became involved in the women's movements in the Palestinian refugee camps and could not devote her life entirely to her boyfriend. She became politically active, led an uprising in which 5000 students participated and joined the trade union. She started organizing working women's committees. Then she discovered that the political parties were only interested in the working women to increase their own power and not to enhance equality among men and women. Her decision to marry the man she loved caused another battle with her parents.

The youngest women in the narratives are starting to pursue professional experience; they are in the midst of trying to find out how to get their priorities right. While to their mothers it seemed obvious that more freedom and more choices would make for more happiness, this freedom – which the younger women also want – has proven a complex and ambiguous acquisition. The problem seems to shift from integrating tradition and modernity, from obtaining more freedom, towards retaining a feeling of 'wholeness' in a world of more choice.

The challenge of choices: moving beyond dichotomized alternatives

• Dolores Rojas Rubio (Mexico), 32, relates how she broke away, time and again, from the usual pattern of a young middle-class woman in Mexico City, supposed to grow up in the parental home, go to school, perhaps aspire to higher education, to marry, settle and have children. But what of those who don't follow this pattern? For Dolores, a conflict with her father made her decide to leave the house while still in school. She remembers her father's final, hurtful words: 'If there's a rotten apple, it is better to get rid of it before the rest go bad.'

Dolores fell in love, married, worked as an engineer; in the evenings she rehearsed for theatre plays, realizing that there were many things that were important to her, including human rights, politics . . . She wanted to involve her husband in all this, but that proved impossible. His reaction was to suggest starting a family, so that she could stay home and look after the children. But Dolores didn't want to be shut up at home. Why should she leave work and give up her own interests? She didn't want to keep quiet about what she thought. She was forced to conclude that she didn't want to be with her husband any more. She was surprised at how used she was to blaming herself, at how anxious and how different she felt when she started to do what she wanted and what she thought was right: 'I felt strange, as though I had suddenly shaken off a heavy weight and was ready to fly.'

• Solange, 31 (The Netherlands), stresses that it is the multiplicity of insecurities which, for the moment, keep her from having children: job insecurity, relationship insecurity and insecurity about the future for new generations.

Her sister Manon, 29, sees the issue rather as a matter of making clear choices and setting one's priorities. And this is something, she feels, that young women of today hesitate to do: they are ambivalent. They make half-choices and then they accuse the men of exploiting them. In fact they are unhappy that their ambivalence does not allow them to flourish fully in both ways. One has to make clear choices and then find a form of living in accordance with those choices.

• Eman Ahmed, 25 (Pakistan), says that until a couple of years ago, getting married was top of her list of priorities. But now she wants to make something of herself before devoting herself to a husband and children. She wants to establish herself as a person and not only as a woman. She wants to be able to support herself if necessary. Once she is married, however, she feels that she will have no qualms about being dependent, emotionally or financially, on her husband.

The female body: marvel and battleground

A close look at the life narratives shows that the women described are not confronted with the same kind of social pressures in all places and at all times. Yet there is at least one striking field of tensions which all women have to deal with: responses to the female body from the social surroundings. The woman's body has long been seen as a marvel, admired for its aesthetic value: one has only to walk the corridor of any art gallery to see that the female form is, and always has been, considered an object of beauty. Yet because of the way certain reactions to it have been institutionalized and sanctioned by religious institutions, cultural practices, political powers and commercial needs, the female body has become a battleground for cultural and religious identity as well as for economic competition.

Examples from the narratives were complemented by other examples emerging during the discussions at the Encounter. They highlight three aspects of this field of tension: male control over the female body; the imposition of patriarchal and commercial perspectives on the female body; and the complicity of religious institutions in targeting the female body for power purposes.

Male control over the female body

• Female circumcision in certain parts of Africa: the story of Sharia (related by Safiatu Singhateh, The Gambia) shows that this mutilation was forced on her, against her will and that of her mother, by an aunt after whom she was named. It was performed in secret by elderly women. The aunt felt that Sharia's outstanding qualities (her school performance, her assertiveness) were obstacles to her future and that the initiation rites would help to 'redirect her ego'.

The aunt's conviction shows that she herself had clearly internalized a societal system that sanctions men's desire to control female sexuality. She believed that, in order for Sharia to be accepted by the community, she must go through this 'rite of passage'.

• Yvonne Deutsch (Israel) tells how, in Israel, the woman's womb is converted into an 'enlisted womb' for the military effort. It is a way of controlling and co-opting the woman's power of reproduction and her close relation to nature (which are at the source of womb envy), underlying many patriarchal structures in the world.

The imposition of patriarchal and commercial perspectives on the female body

• In Ecuador, in the age of Christina Gualinga's forefathers, the Indian peoples saw the first 'civilizers' coming: the celibate Catholic priests, covered from top to toe, some in white robes, others in black. Evangelizing the Indigenous people meant that they had to be 'lifted' from their way of life closely connected with nature: the civilizers apparently considered nature full of sin. The women had to give up their traditional colourful short skirts, as these were perceived as a provocation to men (which men?).

• In India under the Raj, men took to wearing Western suits. At that time, women continued to wear the traditional Indian clothes. Kamala Ganesh feels that the trauma of colonization may have been heavier on men than on women. The men's self-esteem came to be linked to recognition by the colonizer. Imitating the outer signs of belonging to a certain class by dressing their way is a well-known device to enhance self-esteem. By doing so, however, the men may have needed to compensate the frustration of giving up part of their cultural identity. A wife's way of dressing and her general demeanour were perhaps a way for the man to maintain his own tradition. In this sense, women may have served as a source of mental 'security' for men in the struggle for economic security, as well as in the transition to 'modernity'. Similar phenomena are found among fundamentalist Islamic and Christian groups.

• Dolores Rojas Rubio peaks from experience when she claims that in the political sphere in Mexico a woman who is beautiful finds it extremely hard to be recognized as intelligent, let alone as capable of performing political functions.

In Western(ized) culture, sex tourism, beauty contests and sexism in advertising are part of the same perspectives on female capacities and of the same effort to degrade the woman to the level of a bodily weapon on the battleground of commercial interests. The phenomenon of anorexia nervosa among young women who want to respond to the image of the perfect female, promoted by advertisements for cars and toothpaste, is an example of the effectiveness of this weapon, as it even conquers the minds of the young women concerned.

The complicity of religions in targeting the female body for purposes of control or identity

The stories are full of other less striking, but widespread, examples of the way in which religious power is exercised over the female body. The efforts of the Christian churches to impose views on procreation and sexual practice is a very common one. Many women internalized these views up to a point, partly because of a lack of knowledge about birth-control practices and devices (particularly in the more distant past). They rarely challenged this imposition of ideas, but whether they actually agreed with them is another matter.

• Truus Hoeksma, mother of Edith Sizoo (The Netherlands) and nine other children, once sighed and said, 'if God had consulted me just once to ask [whether these "blessings of the Lord" were] also convenient for me . . .'

• In Ireland up until Vatican II (1962) women had to be 'churched' after giving birth to a child. This was a blessing performed by the priest to rescue the woman from the 'unclean' status which the Old Testament proclaimed her to be in. The ceremony enabled her to re-enter the church 'in a state of grace'.
  The church in Ireland has also used the State to impose its views. The two have, according to Ethel Crowley, entered into the unholy alliance of controlling the female body by using the law to forbid divorce and abortion. The strength (and hypocrisy) of this alliance was manifested again at the 1992 referendum which resulted in forbidding divorce, in permitting information on abortion and travelling abroad to get an abortion, but outlawing abortion from taking place in Ireland itself.

But the phenomenon of religions targeting the female body for purposes of identity or power also shows its face in more violent ways. Very recent history has shown how in Algeria, for instance, women walking in the streets have been shot down by extremists of modernity as well as extremists of Muslim fundamentalism, for the simple reason that their bodies were covered either by the veil or by Western clothes.

Similarly, in the former Yugoslavia Serbian soldiers were encouraged to rape Muslim girls, to prevent them from getting married and bearing more children for the Muslim community.

To be fair, it should be noted that it is not exclusively due to men that the female body is targeted for cultural and religious identity or commercial purposes. Patriarchal systems could not operate without the collusion of women: they participate in bolstering militaristic societies, maintain customs like clitoridectomy and collaborate in the use of women's bodies to advertise commercial products. It would be too passive a conception of women to claim that they were simply forced to do so.

To the extent that similar challenges are posed at different times and in different places, they are, however, not always and everywhere equally strongly present. There is a growing diversity of views and practices within families, within societies and between societies. Thanks to new (in particular, technical) possibilities of female control over the female body, together with solidarity through women's movements, male control over the female body is running up against changing views in various places in the world. Similarly, the imposition of patriarchal or commercial perspectives, and the complicity of religious institutions, are not equally strong within different groups of the same society or in different parts of the world.

This variation contributes to the fact that in any given period of time women in different parts of the world and in different sections of society struggle, and have to struggle, on different frontiers. They assign different priorities to the arena in which they want to and can struggle. Their actions vary from straightforward confrontation to peaceful action, humour, irony, dramatization, or hitting hard at economic interests, often using their body in acts of symbolic ju-jitsu.

Symbolic ju-jitsu

• Jaana Airaksinen (Finland) told how a feminist university scholar gave a video presentation on prostitution and representation of the female body in visual arts. She delivered her solid academic speech in lace underwear.

• Durre Sameen Ahmed related an incident from Pakistan, where the alliance between the State and Islam is becoming more and more frightening. A group of mullahs were talking to women about the need to wear the veil in order to protect them against men's eyes, when the women

239

slapped them in the face publicly and (adding insult to injury) subsequently accused the mullahs of molesting them in public.

• Amal Krieshe (Palestine) went on hunger strike when she was denied further schooling. For her, this was a way of saying 'If you do not allow me to take food for the mind, I will not take food for the body.'

• In the Philippines an expert team of the largest gold-mining corporation once came to the Cordillera to see what the prospects would be for opening gold mines. The Indigenous women of the area did not agree with this violation of their ancestral lands. They bared their breasts to scare them away. The team fled from the village.

The challenge of institutionalized religion

Even if, in the same period of time and in many parts of the world, women are struggling on similar frontiers, they may respond differently. This is partly because the tensions, although similar, are not identical. But other factors also play a role, such as differences in the history of generations and, of course, individual personalities. Even responses to similar challenges may differ.

Within the narratives, the most striking example of this appears to be in the responses to institutionalized religion. In almost all the life stories 'religion' is present. In the grandmothers' and mothers' generations there is little open or rebellious questioning of the religious institution as the container of a belief system. The stories also report that the majority of people active in Hindu religious life and in Christian churches are usually women. The daughters' generation in the stories hardly mentioned religion – this is not an issue that profoundly preoccupies them.

The women present at the Encounter expressed many reservations about institutionalized religion. Some had rejected 'religion' altogether, others are still adherents but question certain manifestations of the religion they belong to. Most of them were open to, or explicitly searching for, spirituality as an existential need.

The basic question in this domain of challenges can perhaps be summed up in the formulation of Durre Sameen Ahmed from Pakistan: how to reclaim or sustain, discover or rediscover a spiritual life when the religion itself is being made increasingly repellent all around? In other words, would it be right to think that institutionalized 'religion is a defence against religious experience', as Jung suggested?

• Durre is of the opinion that the choices facing women in the future will be dominated by two issues: the environment and religion. In her understanding, the rise in Islamic fundamentalism in Pakistan is a quest for power by certain groups and perhaps more of a reaction to corrupt and inefficient government than anything else. Islamizing the society means laws and restrictions which are horrific for women. Durre feels that the secularist

Pakistani intelligentsia, claiming to be 'progressive', often have much in common with 'fundamentalists'. Their set ideas about societal issues like 'modernity', 'progress' and 'development' are just as dogmatic as the rejection by their fundamentalist brothers of the same. It is, in her opinion, partly because this Left, 'modern' intelligentsia fails to take seriously the legitimacy of the questions addressed by institutionalized religion that Islamic (and any other kind of religious) fundamentalism gets a chance.

Western science and systems of knowledge have – in recent decades – given women a voice, but they do not allow space for religion. Durre argues that the 'modern', 'rational' man is as unable to deal with complexity and ambivalence as the religious fundamentalist.[1] Durre feels that secularism is not the answer, that religious fundamentalism can only be combatted with good religious arguments from within the religion concerned. All sorts of fundamentalist readings exclude the notion of the feminine. The Koran, however, does not refer to the male as the first human. Eve is not mentioned by name. She is not born out of Adam's rib and is certainly not the temptress responsible for the Fall. In fact it is Adam who was responsible. In its original meanings, Islam liberated women long before Europe. Durre is convinced that it is women – because they are the ones most under threat of fundamentalism – who have to work internationally on a re-visioning of Islam (and any other kind of fundamentalist religion). This is necessary because Islam is not confined to one particular geographical area like Hinduism, but is becoming a worldwide power.

Re-visioning Islam will mean a dual problem for Muslim women: they will not only be besieged from within Islam by mullahs and their political allies, but also from outside it, by their own 'progressive', secular and Christian friends, women and men alike. There is a need to search together for a female spirituality. In Durre's experience, women in the West are yearning for a spirituality which is not prone to a body–mind split.

She recalls that the heart and core of all religious traditions, arrived at after deep contemplation of reality and grounded firmly in the spiritual, points to a certain feminizing of one's inner self in terms of an attitude of receptivity. This is where female/male differences become important, especially as reflected in the metaphors of the body – for example, in the notion of interiority as exemplified by the female.

• For Kamala Ganesh (India), religion is not easily definable. While she sees herself as a Hindu in some respects, advocating such positive aspects as philosophy, spirituality and a sense of anchorage, she dissociates herself from other aspects: caste, inequalities, obscurantism, superstitions and orthodoxy. The typical secular, liberal, Marxist development discourse emphasizes poverty and regards religion as irrelevant, embarrassing or dangerous. For most people in India, however, peaceful religious coexistence is both norm and need. Kamala's efforts at grappling with religious pressures from the family have not led to a total rejection. However, she feels that many Hindu texts are male-oriented and enhance the profound misconceptions of women's inferiority. They may have some emancipatory aspects, but one cannot bend them too much to suit one's inclinations. Why keep trying to reinterpret 'the books'? On the other hand, in the Indian subcontinent, there have been female-centred traditions running in parallel. To Kamala it makes more sense to look for a selective reappropriation of religious values and practice.

• Eliane (Brazil) says that, to her, oppressing forces are found in dominant philosophies, religious dogmas and unagreed ideas, the opportunistic and the cruel. She draws her spiritual force from the dialogue with the ancestors and with some Indigenous leaders (although certainly not all), and from the dignity she received through her mother and grandmother.

• Edith Sizoo (The Netherlands) writes that the Protestant expression of the Christian faith did not induce a spiritual experience in her. The services were centred around cerebral explanations of Bible texts. The head was hardly connected to the body, the heart and nature. She did not find what she was looking for: an expression of the deeper mystery of Life, the kind of spirituality that transcends the limitations of the rational and the cerebral, something that makes one experience one's connectedness with the wholeness of the mystery of the Creation. Religion should create space for people to experience and express that connectedness from within.

She feels that there is a need to explore with women from different cultural (including religious and secular) backgrounds what spirituality means to them and whether one can speak of female spirituality. Women should go beyond complaining about the power devices of male-dominated religious institutions and address the heart of the matter.

• In contrast, Safiatu Singhateh (The Gambia) speaks of the positive effects of Christian ethics with regard to respect for the integrity of the person. In her experience, it is thanks to the presence of the Christian church in Africa that women could be educated and that certain cultural practices which were detrimental to women were (partially) abolished. Her narrative shows, for instance, that women are not in favour of polygamy; and although Christian men are not monogamous, the Church does at least question the behaviour of unfaithful men.

• Esperanza Abellana observes that in the Philippines the Catholic Church reinforces the non-identity of the woman in the family by emphasizing her subordinate role in relation to her husband. In spite of this, many more women than men attend religious services. She wonders whether the women's spirituality is based on their need to implore for added strength and to ally themselves with whoever they believe is the source of power. One of the most heavily attended church activities is the novena to the

Mother of Perpetual Help held every Wednesday. This demonstrates a kind of spirituality that closely interrelates the physical with the spiritual. Do women manifest a world-view that seeks to integrate rather than separate? To what extent is their spirituality a sign of powerlessness and fatalism, or an act of empowerment and strength? A cultural understanding and perception of the women's concept of power would help to shed light on this behaviour.

Esperanza tends to think that religious practice is a woman's way of being 'attuned' with 'forces' bigger than herself. This comes from a realization (or perhaps a basic intuition) that she exists as part of a bigger world – hence her connectedness and relational character, which is affirmed and expressed through her religiosity.

It seems that the common element in all these reactions is in the realm of connections. The women here have a problem with tendencies in religious institutions which lead to separation rather than connection: separation between the institutions, conflicts over interpretations of 'the books' of the religion concerned, disconnections between the mind, the heart and the body.[2] What they are looking for is a spirituality which unites rather than divides.[3] The responses to this common search, however, were not and are not likely to be the same – fortunately – because the diversity of responses is bound to be mutually enriching as long as it is seen as a search for experiencing the wonder that is Life.

SHARING WORDS WITHIN WORLDS

The pitfalls of using a dominant language

Rediscovering the obvious becomes exciting when it proves to be more relevant than one originally thought. Everyone knows that languages are different but that they can be translated into each other. One may also be aware of the limited extent to which a translated word covers the field of associations of the original word. This common knowledge is, however, often not felt to be particularly relevant until one tries to communicate through a foreign language.

The language of communication in the life narratives and during the Encounter itself was English. However, English happened to be the mother tongue of only one of the fifteen authors – and an Irish woman at that, who could claim that English is an imposed language and that her roots are Gaelic.

Because of the necessity of communicating in a 'common' language, the participants almost fell into the trap of reducing their findings to what the common language partially conveyed about ideas more fully expressed in the mother languages. The first warning signal came from the write-up of the stories. These revealed, for instance, that the questions on 'personal integrity' and 'wholeness', as formulated in the guidelines for the narratives, did not work. These concepts were simply left aside and, instead, a series of other (more

243

dynamic) notions emerged, expressing something one can strive for, something that provides sense and a certain personal fulfilment to one's life.

A second signal came during the Encounter, when the participants tried to distil from the stories what the forces were that drove the women concerned in shaping their realities. When trying to find English words for them, the participants discovered that the fields of association of these words, although partially overlapping, were not the same in their own mother tongues.

The truth of this became crystal clear when some of the participants, during the discussion on the meaning of these words, declared that in fact the women in their stories had used words or concepts which could not be translated into English at all. At the request of the group they explained what these untranslatable words meant.

- With regard to the word 'dignity', for instance, which both Christina Gualinga (Ecuador) and Eliane Pontiguara (Brazil) had used in their stories as driving forces for the women of the Indigenous peoples they belong to, Christina said that in her language there is no word for 'dignity' as such. The appropriate word would be *samay*, which means a number of things at the same time: to breathe in a spiritual sense, live in harmony with others, a pure life in relation with nature, but also in relation with the past, the present and the future. To keep the *samay* is at the core of what Quechua women are seeking. It provides them with tranquillity, security, sanity, strength and calmness. They have to oppose everything that destroys the *samay* – things like bulldozers, armies and pollution, which destroy both the spiritual and physical 'breath', since these are one.

- Kamala Ganesh from India added that in her Tamil language there is no word for 'dignity', either. As a driving force she would rather use a Tamil word for 'status', which is close to self-worth and self-respect as opposed to arrogance and false self-esteem.

- Durre Sameen Ahmed from Pakistan mentioned that in Islam one of the myriad names-as-attributes of God is *Al-Rahman*, a central notion for women's motivations. It has multiple meanings and connotations, as is usual in the Arabic languages, where many words are related to at least three different roots. *Al-Rahman* means 'the Gracious One', 'the Merciful One', 'the Beneficent One', 'the Compassionate One', and is etymologically also related to the (feminine) 'womb' and 'sanctuary'.

- For Yvonne Deutsch (Israel), living in a militaristic culture, the notion of *shalom* appeared to be a major driving force, but meaning much more than just 'peace' in the political sense. It also refers to 'wholeness', 'fullness', a way of living in a respectful relationship with the universe.

- Similarly, for the word 'solidarity', it was noted that in Africa this notion has clear connotations which are quite different from the way it is understood in Europe. In Africa the concept is linked to *obligations* which the individual has *vis-à-vis* the group to which she/he belongs, and within which she/he has *face-to-face relationships* (compare the concept of '*amalima*' in Zimbabwe, which refers to 'family units combining their ideas, knowledge, labor and resources in order to improve their standard of living').[4]

In Europe the notion of 'solidarity' is associated with a *free choice* to support other people. Moreover, one does not necessarily know these others in person (for example, solidarity groups supporting Guatemalan indigenas, or economic solidarity through the State social security system, which is *anonymous*).

• Esperanza Abellana (Philippines) explained that, when asking women in the Philippines what their aspirations are, what they strive for, they would answer '*makaginhawa*', which means 'to be able to breathe'. Beneath this literal meaning, the word implies a longing for well-being or relief from a difficult situation – economic needs of the family, education for the children – but always related to seeing the children develop a harmonious relationship with God, the community and the family. The woman's identity is not separate from her experiences of fulfilment, struggle and crisis. This is the context of her 'wholeness' as she responds to personal challenges in relation to family and community. Like the wave in the ocean that appears visible and distinct as a wave form for a moment, the woman's individual identity is always in constant movement: rising up and submerging with the shared identity of the people she is in relation with. The next moment the ocean wave blends itself in ever-widening ripples with the foam and the waters. As she copes with all the expectations and the responsibilities, she feels the pain of being submerged, of losing breath. Paradoxically, she emerges with a *lakas ng loob* (a strong inner self), because she chooses or has the *buot* (consciousness and wilfulness) to respond to the life needs of her family and her society.

There was great value in elaborating on the content of these and other notions. On the one hand, it highlighted in a concrete way the great extent to which one is indeed prisoner of one's own language; and even more so of a chosen means of communication in international relations, in this case the English language. On the other hand, it demonstrated that a conscious attempt to cross the boundaries of that prison does indeed open the window to complementary perspectives on life.

Examining the question 'what's in a word' also underlines the fact that this type of exploration can bring to the fore a great variety of motivating factors, as well as their actual consequences for women's behaviour in dealing with their everyday realities. A much more systematic effort to develop this method may provide precious basic material for culture-sensitive approaches to supporting women in various parts of the world, particularly within the framework of development policies and practices.

NOTES

1 See the author's introduction in Ahmed (1994).

2 The idea of 'embodiment' is also taken up by Rosi Braidotti, albeit not explicitly in relation to spirituality. In the section entitled 'Repossessing Bodily Space: A Timely Project' of her chapter, 'Radical Philosophies of Sexual Difference' (Braidotti, 1991: Chapter 8), she states that 'the emphasis on the body coincides with the post-Nietzschean appeal to overcome the classical mind–body dualism in order to think anew about the structures of human subjectivity'. Referring to the French-based movement of the *écriture féminine*, in which the question of the body has emerged as central to radical feminist philosophies of the subject, she notes that the body

> cannot be reduced to the biological, nor can it be confined to social conditioning. In a new form of 'corporeal materialism', the body is seen as an interface, a threshold, a field of intersection of material and symbolic forces; it is a surface where multiple codes of power and knowledge are inscribed; it is a construction that transforms and capitalises on energies of a heterogeneous and discontinuous nature.
>
> (Braidotti, 1991: 219)

3 Charlene Spretnak (1993) takes a stand against deconstructive postmodernism as leading to denial of meaning and thus to meaninglessness, which she sees as the great pitfall of the modern age. Instead, she invites us to rethink and, more importantly, to experience the relevance of the core teachings and practices of the great spiritual wisdom traditions for everyday life, including economic or political issues: 'To partake of the wisdom traditions we need to explore possibilities across parochial boundaries and to appreciate core spiritual insights independently of the institutional religions that may have grown up around them.'

  Seeking 'illumination' of central issues of our time, she examines the teachings of the Buddha (in the areas of 'mind, perception and mental suffering'), the spiritual practices of Native Peoples (in the area of 'intimate connection with the rest of the natural world'), the contemporary renewal of Goddess Spirituality (in the area of 'consciousness of the body as intricately embedded in a relational web') and the core teachings of the Semitic religions, Judaism, Christianity, Islam (in the area of 'social ethics as an expression of our comprehension of the divine oneness').

4 Florence Mafeking, representing the Organization of Rural Associations for Progress (ORAP), reported at the El Taller Workshop on Gender and Development (Tunisia) that this Zimbabwean Indigenous grassroots movement for self-reliance mobilizes people through *amalima*, the concept of family units working together. This traditional custom which has recently been jeopardized by individualism, is being revived by ORAP.

REFERENCES

Ahmed, D.S. (1994) *Masculinity, Rationality and Religion: A Feminist Perspective*, Lahore, Pakistan: ASR Publications.

Braidotti, R. (1991) *Patterns of Dissonance*, trans. Elisabeth Guild, Cambridge: Polity Press.

Katz, C. and Monk, J. (eds) (1993) *Full Circles, Geographies of Women Over the Life Course*, London: Routledge.

Massey, D. (1994) *Space, Place and Gender*, Minneapolis: University of Minnesota Press, pp. 154–155.

Spretnak, C. (1993) *States of Grace*, New York: HarperCollins.

POSTSCRIPT

Shanti George

The Encounter as experienced from within

I have taken part in many projects and conferences, but the Network Cultures project on 'Women's Ways of Shaping Their Realities' has involved somewhat unconventional participation.

The first unusual feature was being selected to take part in the project on the basis of my familiarity with certain everyday realities, with contexts rather than with texts. (I am usually invited to join projects because I have written such and such, or am familiar with particular literature.) The same applied to other members of the project's group, so that discussing what we had lived was at least as important as debating what we had studied. This was a departure from what I am used to, namely the 'professionalism' that stipulates that one keeps oneself out of the picture.

This emphasis on what is usually seen as 'informal' (and outside the formal world of most study projects) carried over to the meetings of the participants, where we met as multifaceted individuals and not as unidimensional 'resource persons'. Another unusual feature of these meetings was the very loose agenda for discussion, that was made even looser by what many would see as digression and meandering. At first I mentally gave up expecting that the main issues would be covered in the time available, but somehow they generally were (sometimes in a final inspired half-hour of 'Let's finish what we've come here to do') – so that eventually one came away satisfied, as well as relaxed from the meandering!

As with other participants, the idea of writing about my own experience seemed attractive, a bridge between context and text. After several years of social science training, I turned some of its searchlights on to myself instead of others. The process of writing about my own family eroded some barriers: it authenticated my professional training against the material most familiar to me, at the same time it provoked wider and stronger interest among my family circle than any of my previous writings.

I was simultaneously 'revealing' myself to the new circle of project participants. By reading their papers, I became intimate with the lives of women I had mostly yet to meet – and so when we did meet, in some important ways we were no longer strangers. I did not know what they would look like or sound like, but I knew some significant things about them 'under the skin'. This was in contrast to other projects and workshops, where personal details were never 'up front'.

The Encounter was characterized by relatively free-flowing discussion rather than by structured discussion of papers. I found that I could test my perceptions and interpretations among a circle of experienced and interesting women, who – at the same time that they were fascinatingly diverse – shared some major concerns about women's lives. On the agenda were the sorts of things that at other workshops I had discussed briefly and often wryly with co-participant women, but only sometimes and then only in the interstices of the formal proceedings.

Very personally, I was encouraged to bring my 7-month-old baby along to the Encounter, with my husband to look after her. He brought the baby to me to breastfeed at the sessions, once even when I was chairing the session. At that point, there was no other way I could have attended a meeting (and there were probably few other workshops that I could have attended in this way).

To summarize, this was one project and one encounter in which I participated as a person with many dimensions, and with both a personal life and a professional life that could be integrated in the project work. Of course, many projects and workshops by their very nature could not function like this.

The project's manner of studying women's ways of shaping their realities reflected what to many is the ideal-typical or stereotypical manner in which women have traditionally operated, usually described as informal rather than formal, spiralling rather than linear, elliptical rather than direct, expressive rather than instrumental, holistic rather than fragmented, and collective rather than individualistic. While I would not treat these qualities as representing some 'essence' of the female, I would agree that they have tended to characterize women's ways in a wide range of cultures. Thus, the project achieved an unusual congruence between its subject matter and its procedures.

INDEX